Flying VFR in Marginal Weather

Second Edition

Paul Garrison,
Revised by Norval Kennedy

TAB BOOKS Inc.
Blue Ridge Summit, PA 17214

SECOND EDITION

FIRST PRINTING

Copyright © 1987 by TAB BOOKS Inc.

Printed in the United States of America

Reproduction or publication of the content in any manner, without express permission of the publisher, is prohibited. No liability is assumed with respect to the use of the information herein.

Library of Congress Cataloging in Publication Data

Garrison, Paul.
   Flying VFR in marginal weather.

   Includes index.
   1. Airplanes—Piloting. 2. Meteorology in
aeronautics. I. Kennedy, Norval. II. Title.
TL710.G34   1987   629.132'5214   87-7133
ISBN 0-8306-9416-1
ISBN 0-8306-2416-3 (pbk.)

Questions regarding the content of this book
should be addressed to:

   Reader Inquiry Branch
   Editorial Department
   TAB BOOKS Inc.
   P.O. Box 40
   Blue Ridge Summit, PA 17214

Contents

Preface

This book is not intended to encourage pilots to fly into marginal weather conditions or to take silly chances of any kind. Its purpose, rather, is to demonstrate how such situations can be dealt with when a pilot has inadvertently come face to face with conditions which are beyond the limits of his experience or capabilities or go beyond the capabilities of the aircraft involved.

All of the examples used in the following pages are based on actual incidents. Any names and aircraft identifications used are fictitious in order to protect those involved. Quite frequently the solutions to a particular problem involve actions by the pilot which may be contrary to the Federal Aviation Regulations (FARs), and must be looked at as emergency measures. It goes without saying that pilots should, when at all possible, operate within the rules and regulations set forth in the FARs. Still, when faced by what appears to be at the time an unavoidable choice, safe is to be selected in preference to legal.

With weather continuing to be the primary cause of accidents among general aviation pilots in general and Visual Flight Rules (VFR) pilots in particular, most of the case histories used in the book are weather-related incidents. But we have also included some in which the weather, as such, was not the primary cause of the developing danger.

An uninformed reader may get the impression from what follows that flying light aircraft involves a never-ending series of hair-raising situations in which the pilot and his passengers face nearly insurmountable dangers. That, in fact, is not the case. The average and reasonably proficient pilot can spend thousands of hours at the controls of light aircraft without encountering any serious emergencies. If he is careful and resists the temptation to take unnecessary chances, it is unlikely that he will ever be faced with the need to make a life-and-death decision. But flying, like most other human activities, involves a certain calculated risk, and it is in an effort to minimize that risk factor that this book has been compiled.

Paul Garrison

Introduction

"It is unavoidable in a book of this type that we must talk frequently about actions which are not only illegal but also dangerous to the occupants of the aircraft involved and to other innocent aircraft which may be operating quite legally in the same airspace.

"It can't be emphasized too frequently or strongly that such illegal operations in instrument conditions should be considered *only as a last resort*.

"Any rogue pilot who habitually ignores the FARs and blithely operates on instruments without bothering to obtain an ATC clearance is, in fact, a menace to aviation in general and to himself in particular."

Chapter 18

Considering that, as you read this revised and updated edition of *Flying VFR In Marginal Weather*, please remember it has been published purely for the enlightenment of the reader.

Certified flight instructors, the FARs, pilot's operating handbook, or various updated communications by entities mentioned in this book, and other entities, may alter or affect the information published.

No individual, organization or company associated with this book assumes any liability arising out of reliance upon material contained herein.

Any navigation charts or other material resembling navigation charts in this book are supplied as reference material only; they are not for navigation purposes.

Norval Kennedy

Chapter 1

What Is Marginal Weather?

It is obvious that in order to avoid flying into marginal weather situations, we must first be able to detect them. But there are no hard and fast rules. The term may mean one thing to a pilot with limited experience and something entirely different to one who has been flying for years. Most generally it is thought, of as a condition of lowering ceilings and deteriorating visibility. But marginal conditions might also involve strong winds, turbulence, precipitation, extremes of heat or cold, excessive flight altitude without on-board oxygen, smoke, haze, smog or icing under adverse temperature and humidity conditions. Let's take one at a time.

Ceiling

The minimums for legal flight in controlled airspace call for 1,000 feet between the ground and the base of the clouds. In uncontrolled airspace there is no ceiling minimum—the rules simply require the pilot to stay clear of clouds. Technically a ceiling is the height, agl, of the lowest layer of clouds described as either broken or overcast, but not classified as thin, obscuration or partial obscuration. Thus a VFR pilot may legally obtain takeoff clearance at a controlled airport even if there are scattered clouds with bases below 1,000 feet agl, or if the conditions are described as sky partially obscured.

Once he has left the airport traffic area it is up to him to decide whether the conditions along his route of flight continue to be adequate for safe conduct of the remainder of his flight. In flat country it may be perfectly possible and quite safe to fly long distances under a 1,000-foot ceiling, always assuming that the pilot is aware of broadcast towers or other man-made obstacles and that he remembers that whenever he is about to overfly an area which is located within a five-mile radius of a controlled airport, he must contact the tower for permission to pass through the airport traffic area. This type of flying is not particularly comfortable and frequently results in pilots getting themselves into a box from which they find it hard to extricate themselves. But more about that later.

Visibility

Visibility, as used in aviation, is the greatest horizontal distance at which an observer can identify prominent objects with the naked eye. In controlled airspace the minimum visibility for legal VFR flight is three statute miles. In uncontrolled airspace it is one mile. Judging visibility on the ground is usually accomplished by looking at a landmark, the distance of which is known. Once in the air, on the other hand, judging visibility becomes a strictly subjective matter. The pilot looks out of the window of his airplane and simply guesses as to how much visibility he has. Much of this is necessarily based on how far ahead he can see the ground since in the air there is nothing to look at, making it impossible to have any clear idea of how far one can see.

This presents no problem when it's reasonably clear. The difficulty arises when we are flying in haze conditions. Relatively thick layers of haze tend to cover large portions of the country much of the time (Fig. 1-1). What usually happens is that we can see the ground below without too much difficulty, but we don't have the faintest notion of how far we would be able to see another airplane if there was one at our altitude. In most instances we are under the impression that the visibility laterally is less than it actually is, and when we finally do spot another airplane, we are surprised to find that we can see him quite clearly in all that soup even though, judging by his size, he is quite a distance away. Flying in such haze conditions, while technically VFR, more often than not requires that the pilot be able to control his airplane primarily by reference to his instruments because the horizon may be indistinct or totally invisible.

Wind

Winds of average velocities, say less than 20 knots, present

Fig. 1-1. VFR pilot takes off into low-visibility haze conditions.

no particular problem except that they influence our ground speed and may, unless we pay close attention, blow us off our desired course. And for tailwheel aircraft they may cause problems during takeoff and landing when the available runway doesn't head straight into the wind. Stronger winds usually produce uncomfortable turbulence and, especially in mountainous areas, severe up- and down-drafts. Winds, especially head winds, must be figured in terms of percentages of the cruising speed of the airplane. A slow airplane is much more seriously affected by head winds than is a faster one. Thus, an airplane with a true airspeed of 100 knots, fighting a 25-knot head wind will be moving across the ground at 75 knots or only three quarters of his no-wind speed, meaning that the distance between refueling points has to be reduced by 25 percent. On the other hand, a 200-knot airplane, fighting the same wind component, will be losing only 12.5 percent of both speed and range. The basic rule during severe wind conditions is to pay closer than normal attention to navigation and to be prepared to change plans when necessary.

Turbulence

Most turbulence normally encountered is simply an uncomfortable nuisance. It can be caused by gusty winds, by wind blowing over mountains or other ground-based obstructions, by rising warm air or the atmospheric activity in the vicinity of thunderstorms. Though officially classified as light, moderate, severe and extreme,

3

judging its severity depends on the pilot's susceptibility to the resulting discomfort, and on the size, type, weight and speed of the airplane involved. Airplanes are built to be able to withstand huge amounts of turbulence and the danger associated with severe turbulence is usually not one of structural damage to the airplane, but rather the chance of the pilot losing control. In his effort to regain control, he may inadvertently cause stresses which exceed the limits of the aircraft structure. As a general rule it is advisable to reduce the speed of the aircraft somewhat and not to constantly try and correct for up- or down-drafts. Let the airplane bounce around and try to limit control inputs to those necessary to stay on course and maintain adequate obstruction-clearance altitude.

Most of the time, conditions near the ground tend to be more turbulent than at higher altitudes, although this is not always the case. The worst turbulence is that associated with thunderstorms and no aircraft should ever intentionally be flown into a thunderstorm. We'll talk about that later in greater detail.

Precipitation

Precipitation comes in a variety of forms, some quite harmless, others to be avoided at all cost.

Rain. Light drizzle or steady rain is of no particular consequence except that it may reduce forward visibility to a degree and reveal the apparently unavoidable fact that most light-aircraft cockpits tend to eventually leak. In aircraft with carbureted engines it may be a good idea to apply carburetor heat because moisture may produce carburetor icing, no matter what the outside air temperature is.

Really heavy downpours are a different story. They may effectively reduce the forward visibility to a fraction of a mile, not to mention the unpleasant sensation of the noise which tends to sound like machine guns firing at the windshield. For aircraft with carbureted engines the same applies as mentioned above. Otherwise the aircraft itself won't feel any ill effects from such heavy rain, though paint has been known to peel off in spots. If at all possible, it is always the better part of valor to detour around such showers.

If the downpour comes out of the bottom of a thunderstorm, it is strictly a place to stay far away from. Not only is the area of precipitation the most likely to contain repeated lightning strikes, it will probably also be extremely turbulent. There is always the

4

possibility that some of the rain may turn to hail, large enough to severely damage the aircraft.

Freezing Rain. In the winter we often run into situations where the air aloft is comfortably above freezing, but the air below hovers around the 32 °F level. If rain starts falling from above it may turn into freezing rain when it hits the colder air below, or it may simply turn to ice when it hits the metal surfaces of the aircraft. Either way it can coat an airplane with a sufficiently thick layer of ice to impair its ability to carry all that additional weight. The stall speed rises rapidly while the TAS is gradually reduced, eventually making a descent and landing inevitable. Depending on the degree of precipitation, this sort of thing can develop within minutes. Therefore, whenever freezing rain is even considered a possibility, it is a lot wiser to abandon the flight than to wait and see in the hopes that it might not develop after all.

Hail. Hail is never anything to take lightly. It is produced by supercooled air within thunderstorms but often falls several miles from the storm itself, usually out of the overhanging anvil clouds. Such hail has been known to attain golfball size and when it hits the leading edges of an aircraft wing or the fuselage, it can dent the surfaces to a point that airworthiness is reduced to nil, and subsequent repairs can run into astronomical figures. At the first indication of hail a hasty retreat is the only sensible action.

Snow. There is dry snow and wet snow. Dry snow presents no danger to the aircraft, but the visibility may be reduced to near zero and, especially at night, the visual effect of the stuff coming at us is extremely disconcerting and may eventually produce a degree of vertigo. Wet snow is likely to adhere to the aircraft and freeze into ice, similar to freezing rain. In addition, it may cover the windshield to a point which is beyond the capability of the defroster to deal with, totally robbing the pilot of all forward visibility.

When there is snow on the ground it affects takeoffs and landings. A thin layer of snow, while reducing the rate of acceleration during takeoff run somewhat, is not a serious problem. A heavy layer, say three or more inches, may double or even triple the distance needed to attain flying speed. During landing any layer of snow, no matter how thin, reduces the effectiveness of the brakes. A thick layer of snow, especially wet snow, may cause such sudden deceleration upon touchdown that the aircraft may damage the nosewheel or even nose over. If a landing must be made in relatively deep snow, it should be a short-field landing with the nosewheel held way up just as long as possible.

Temperature

Extreme heat itself is rarely a problem at any reasonable flight altitude. The problem is indirect. The sun heats certain portions of the ground more effectively than others. The resulting thermals, columns of warm air rising rapidly, are what makes it possible for the sailplane pilot to often stay aloft for hours, but they simply mean unpleasantly turbulent conditions for the pilot flying a powered aircraft. If there is any moisture in the air, each such thermal is likely to be topped by a small cumulus cloud and the turbulence will extend upward to the base of those clouds while the air will be smooth above.

On the ground, summer heat can raise the temperature in a parked aircraft to 180 degrees or more, and such heat tends to eventually damage the avionics equipment in the airplane. Leaving the vents open helps some, as does covering the instrument panel with some type of heat and light reflecting material.

Extreme cold also presents few problems in flight as most aircraft are equipped with fairly decent heaters. As a matter of fact, reciprocating engines work better and more economically when the outside air temperature is low.

It's on the ground where cold really gets us down. Engines are hard to start or won't start at all without preheating. Oleo struts are suddenly low and may have to be pumped up. Preflighting the airplane becomes a noxious chore, and taxiways and runways may be covered with patches of ice which can make taxiing, takeoffs and landings hazardous.

Chapter 2

VFR on Top

One of the all-time favorite means by which VFR pilots get themselves into trouble is to get caught on top with no way down. The reason is simple and quite easy to understand. You take off from an airport with a broken ceiling at 2,000 or 3,000 feet, and it's bumpy and ugly and you get bounced around, yet above and between the broken clouds there is that gorgeous blue sky virtually begging you to shove the throttle full in, haul back on the yoke and start climbing.

And that's what you do. You weave back and forth to stay clear of the clouds which soon are rushing by you on either side, and then it's suddenly smooth and a few minutes later you're above all that mess. You sit back, take a deep breath, relax and say to yourself, "This is what flying is all about." It's marvelous, comfortable, cool (Fig. 2-1). The true airspeed is higher than it was down below and the fuel consumption is lower with the mixture adjusted to just barely on the rich side of peak.

And so you're flying along, fat and happy, while the breaks in the undercast gradually get smaller and smaller, until after awhile there aren't any breaks down there. Okay, as long as there is plenty of fuel in the tanks, you figure that there is bound to be a place somewhere between here and your destination where you'll be able to get down.

Fig. 2-1. It's smooth and sunny above the weather.

You Spot Some Buildups

Now you spot some higher buildups sticking up out of the top of the undercast ahead and you notice that the distance between your flight altitude and the top of the undercast is gradually diminishing (Fig. 2-2). But you've still got quite a few thousand feet

Fig. 2-2. Now you spot some higher buildups sticking out of the undercast ahead.

between the level at which you're flying right now and the altitudes at which you might have to start worrying about oxygen, so you trim the nose up a bit and maybe increase the manifold pressure if you're not already operating at full throttle. You watch the altimeter move slowly upward while you manage to continue to stay safely above those white billows below.

By now it's been a couple of hours since you last had a glimpse of the ground below. You've been navigating by radio and may have found that despite the increase in true airspeed, your ground speed has deteriorated somewhat, indicating that you must be fighting a fair amount of headwind at these higher altitudes. The fuel gauges show half for both tanks, so fuel is not yet a problem. Still, there is that little voice in the back of your head that keeps reminding you that one of these days you're going to have to think in terms of getting down from up there. You start to pay closer attention to the weather sequences and find that most stations along your route of flight are reporting solid overcast conditions, though the ceilings are ample and there is no visibility restriction to speak of.

Descent Options

You're now beginning to think about the available options. Let's say you're flying at 10,500 feet msl and you're somewhere over western Kansas where the terrain is level with no mountains to worry about and the ground elevation, according to your Sectional is around 2,000 feet. According to the last weather reports you heard, the ceilings average 4,000 feet which puts them at 6,000 feet msl. Your best guess is that the undercast is about 1,000 feet below you which puts it at 9,000 feet or so, indicating that the clouds are about 3,000 feet thick. At a rate of descent of 500 fpm this means that it will take about six minutes to get through if that turns out to be the only way.

Well, it's still too early to worry about it. You still have sufficient fuel for another 200 or so miles and you figure that there is bound to be a hole along there somewhere. As a matter of fact, the last weather sequence did include some reports of broken conditions somewhere about 150 miles to the left of your intended course. Should you turn left and suffer the consequences of that huge detour, or should you take a chance, stay on course and hope for the best? Maybe if you'd climb still higher you'll be able to spot some of those dark patches in the clouds below which usually indicate thin spots or actual breaks. Why not? Let's try it and see.

When level at 12,500 you do believe that what you see ahead

and somewhat to the right of your course is a patch of darker clouds and you head toward it to determine whether there are some breaks. By doing this you effectively eliminate the escape route to that area of reported broken conditions which will soon be beyond the available range with the remaining fuel. It seems to be taking forever to reach that dark patch of clouds you're aiming at and you find that your eyes move with increasing frequency to the fuel gauges which are now hovering around the quarter mark, meaning that it won't be very much longer until the choice of action is reduced to one, namely, get down.

You're In Luck

Now, assuming that those dark patches in the clouds actually did turn out to be what you thought they might, you're in luck. Let's say it isn't an actual hole, but at least a thin spot which gives you the impression that you can see through the cloud deck. Is what you see actually the ground or is it just the top of another layer of clouds in the grey murk below? You may want to circle for a moment to make sure. Any straight line, no matter how vague, would indicate something man-made, in other words, the ground. Okay, there it is, something straight, probably a road, though you can't be sure. Still, it indicates that you'll be able to get down without losing visual contact with the ground for any length of time. You briefly consider that technically what you're about to do is not legal because you won't be maintaining the required distance from clouds, but forget that for now. Right now getting down is more important than worrying about the FARs.

If you're still thinking clearly and are planning ahead, you'll now check your nav receiver for your exact position with reference to an airport which you have to locate once you're below the clouds. You'll have to assume that your descent will, most likely, be a shallow spiral in order to stay within the confines of that patch of thin clouds and that you'll therefore be more or less at the same location below the clouds which you are now at above them. These preparations are extremely important because once below the clouds and relatively close to the ground you may find that you've lost signals from the VORs you are currently using. Unless you know which way to fly and what landmarks to look for, you could end up running out of fuel while looking for the airport.

Now, knowing in which direction the airport is located and how far away it is, you're ready to descend. Regardless of how much or how little you are able to see of the ground, you'll have to be prepared to fly that decent by reference to the instruments alone, because the turning radius of the aircraft is likely to be such that you'll be in and out of clouds all the way down. So you throttle back, possibly apply carburetor heat just to play it safe and trim the airplane to a comfortable 500-fpm rate of descent. You fly a shallow bank in order not to lose sight of your hole and try to maintain the rate of bank at a steady angle. First all this is easy. It will take several minutes until you get to the tops of the clouds and everything is working just fine. But then, quite suddenly, everything changes. Clouds stream by your window at an incredible speed and even though the actual speed of the airplane has remained unchanged, you have a feeling as if you were suddenly being propelled at some incredible jet speed through the mess which envelops you on all sides. Forget about looking for the ground. Right now the only important thing to watch is the artificial horizon, the airspeed indicator, the vertical-speed indicator and occasionally maybe the altimeter.

Again, it just seems to be taking an incredibly long time. Won't those clouds ever end? And then there it is, the ground. A road, fields, a house. You're out of the clouds. You level off at 4,500 feet which puts you some 2,500 feet above the terrain. You check the directional gyro against the magnetic compass to make sure that it's still pointing in the right direction and you take up the heading which you had determined earlier will get you to the airport.

Okay, in this case you were lucky and you made it all in one piece. But what if there hadn't been that convenient hole or thin spot in the overcast? Or, this being your first time, so to speak, on instruments, you'd have gotten spooked and lost control over the aircraft while circling down? Let's briefly analyze the various possibilities.

The Overcast Is Solid

The overcast is solid and there is no hole. In this case the worst possible thing to do would be to keep on flying and hoping until the fuel-gauge needles hover on the empty line. As long as you've got fuel, you've also got control but once the fuel is gone, you not only have to come down whether you want to or not, you don't even have the opportunity to look for a place to land once you're in the clear below the clouds. In other words, don't wait until the last possible minute. Make a decision when there is still plenty of fuel to spare. And even then, be prepared to run one tank dry when there are still 10 or so gallons in the other. It's silly to keep switching tanks until both are practically empty and neither can be counted on to take you that final distance to the airport.

So with the fuel getting low, but not yet critical, you pick an airport at which you want to land. You may want to pick one that is uncontrolled since your descent through the clouds will be illegal, though it is unlikely that the controllers in the tower would actually see you break out of the bottom of the overcast, even if this should take place right over the airport. Realizing that there is always the slight possibility that you could run into some IFR traffic while in the clouds, you now have to decide whether you want to talk to air-traffic control or whether you prefer to simply bore a hole in the clouds and hope for the best.

Even without an instrument rating, the right thing to do is to contact the nearest FSS and to tell them that you want to file IFR. If you sound as if you know what you're doing, they are not going to ask you whether you've got an instrument ticket. They'll simply ask you to go ahead with your flight plan. But you cannot file an IFR flight plan simply from VFR above to VFR below. You have got to include a destination. Thus you will have to give your current position and altitude, a destination airport, regardless of whether that is the airport at which you actually want to land, the estimated time at which you'd be arriving at that airport based on the estimated ground speed of the airplane you are flying and so on. The FSS will then ask you to stand by while they call ATC on the phone and file the flight plan for you. After a while, and sometimes this can take 10 minutes or more, they get back to you with "ATC clears Cessna one-two-three-four-Romeo to the Middletown Airport direct, descend to and maintain 6,000. Contact Kansas City Center on 134.9, over." You are now expected to read that clearance back and not being proficient at instrument procedures, you're likely to find that you've missed half of it. The sensible thing to do is to have a piece of paper and a pencil handy and to jot down such data as altitudes and frequencies. That will make the readback easier and you've got it handy for later reference.

Once the clearance has been read back correctly, you contact ATC on the center frequency, simply telling them: "Kansas City Center, Cessna one-two-three-four-Romeo at one-one thousand, descending," and they come back with: "Cessna Three-Four-Romeo, radar contact. Report reaching 6,000." This means that they have you on radar and that they want you to tell them when you have reached the 6,000-foot altitude. The clearance would probably also have included a specific transponder code and ATC may ask you to ident in order to make sure that they are looking at the right airplane.

Still in The Clouds

The problem with such an IFR clearance is that the altitude to which you have been cleared may not actually be below the clouds. It could just as easily leave you level in the clouds. If that happens, you contact ATC when reaching your altitude and say: "Kansas City Center, Cessna three-four-Romeo level at six, requesting four." They may give it to you, assuming that the requested altitude is at or above the minimum en route altitude (MEA) for the area in which you are flying. If they won't clear you to the lower altitude but you feel uncomfortable continuing in the clouds, you can call them back and tell them that you're cancelling IFR. They then assume that you're in VFR conditions even though you're not, and they'll stop worrying about you. You can now drop down to VFR conditions below the overcast, being reasonably sure that you're not going to run into anybody because there is no IFR traffic below the published MEA.

The trouble with filing instruments like that is that ATC expects you to be proficient on instruments and capable of following the controller's directions. All of this doesn't sound too complicated when we're sitting on the ground and thinking about it. But up in the air, while fighting some turbulence, worrying about the remaining fuel and wondering what it might be that ATC is going to tell us to do, it tends to get a bit hairy.

The other option, short of simply coming down illegally without talking to anyone, is to call the nearest FSS and simply confess your predicament. Though in the end this may result in having your license suspended for three months or so, they will at least do everything in their power to get you down safely and to keep you away from any other traffic. Most probably, if there is a controlled airport in the area, they will direct you to the airport and then give you a surveillance radar approach because it is the easiest to fly for someone with no instrument experience. It consists of a man on the ground telling you which way to fly, which way to turn and when to descend. Since he knows only where you are laterally, but not vertically, you are expected to keep him informed as you pass through the various altitudes. As long as you keep your cool this type of approach is a lead-pipe cinch.

One word about controlling an aircraft by reference to instruments alone. Though it has been said and written hundreds of times, it bears repeating. Our sense of balance is completely unreliable once visual reference to a fixed object such as the ground or the horizon is lost. When the airplane is in a bank for just a few moments, the inner ear which controls our sense of balance will try to convince us that we are flying straight and level. Then if we actually level off, it will try to tell us that we are banking in the opposite direction. We simply must force ourselves to completely ignore the so-called seat-of-the-pants sensations and must place complete trust in the instruments. The primary instrument used for this purpose is the artificial horizon. It tells us the position of the wings relative to the horizontal plane. Then the airspeed indicator is used because a sudden reduction in speed will mean that we are climbing and might be approaching a stall, while an increase in speed indicates that we are diving at too great a rate. The vertical speed indicator (VSI) is also helpful, but its reactions tend to lag, causing us to overcontrol. The turn-and-bank indicator is useful in maintaining a steady angle of bank, although this can also be accomplished using the indications of the artificial horizon.

Whenever we have someone with us in the airplane who can look out of the window and watch for traffic, it is a good idea to practice flying straight and level, banking, climbing and descending by instruments alone. At least then, if we ever have to do it, we know what to expect. While precision flying is of no particular importance when we're VFR, it becomes an absolute necessity the moment we find ourselves in IFR conditions and it is, at times, amazing how difficult it can prove to be.

Chapter 3

Fronts

An understanding of how and why fronts are formed, how they move and the type of weather usually associated with them is of lesser importance to the VFR pilot than to the one who habitually operates in instrument conditions because the VFR pilot is supposed to fly where he can see the weather. Still, it doesn't hurt to have at least some idea of how all this weather business functions.

Cold and Warm Fronts

If the air everywhere were of the same temperature and moisture content, then there would be no wind and, in turn, no weather. But the atmosphere around us is punctured by low-pressure and high-pressure areas around which the air circulates in predictable directions. Each low-pressure area is normally associated with two so-called fronts, a *cold front* and a *warm front*.

A front consists of a sloping transition zone between two air masses of differing temperature and moisture content. A warm front is formed when warm air replaces cold air at ground level and slopes upward over the cold wedge. When cold air displaces warm air at ground level it is referred to as a cold front. The slope of an active cold front is usually significantly greater than that of a warm front. In practice this means that high clouds in advance of a warm front may appear as far as many hundred miles ahead of the front itself, while advance clouds indicating a cold front appear at only relatively short distances ahead of the front. But cold fronts tend to move faster than warm fronts and the fast-moving nose of the cold front forces warm air upward quite violently, resulting in strong winds and convective squalls. In short, cold fronts are usually associated with quite violent weather, winds and turbulence, while warm fronts are fairly calm with light winds and little turbulence. The other difference is that warm fronts often cover large areas and may stay around for a long time, while cold fronts are smaller in area and move on quickly. In either case, depending on the moisture content of the air, the extent of the cloud formations may vary horizontally and especially vertically. With ample moisture available, cold fronts may build cumulonimbus clouds which can attain altitudes of 30,000 feet or more while excessive moisture in association with a warm front will increase the horizontal rather than vertical formation of clouds.

Rain Areas

Both types of fronts will usually involve rain. The rain area of a warm front can extend for hundreds of miles while the rain associated with a cold front may be heavier, but is usually restricted to much smaller areas. Figures 3-1 and 3-2 show the typical profiles of such fronts, while Fig. 3-3 illustrates the typical position of fronts with relation to the low in the early, middle and latter stages. Since the airflow around highs and lows is consistently the same, clockwise (anti-cyclone) around a high and counter-clockwise (cyclone) around a low (Fig. 3-4), knowing the position of high- and

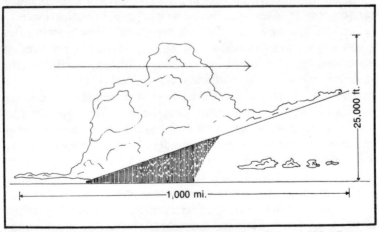

Fig. 3-1. The effects of a warm front may extend as far as 1,000 miles.

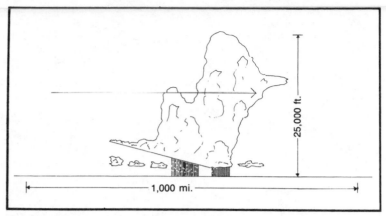

Fig. 3-2. Cold fronts are usually smaller in area, but more violent.

low-pressure areas can be helpful in planning a course which will take advantage of favorable winds.

Weather Charts

The VFR pilot who takes the trouble to become familiar with weather charts and learns what all those hieroglyphics mean, and who then takes the additional trouble to go to the nearest FSS to look at the current weather charts, will start on his cross-country flight knowing where to expect weather and how to avoid the worst headwinds. Figure 3-5 is a typical weather chart covering all of the contiguous United States.

Now let's say that a pilot wants to fly from El Paso, Texas, to Pierre, South Dakota. The straight line would take him right through the middle of a low-pressure system with quartering headwinds along most of his route. By crossing over to the other side of the cold front which lies more or less along the border between Texas and New Mexico and which, according to the chart, does not have any significant precipitation associated with it, he can take advantage of strong tailwinds (30 to 35 knots) all the way into Nebraska.

There he'll encounter that same cold front once more, and a small area of active precipitation is indicated between North Platte and Pierre, but he will probably be able to fly around it and the winds will still be helping him rather than holding him back. In terms of distance, the straightline route is approximately 800 nm, while the detour around the low would add another 125 nm. Assuming that the airplane being flown cruises at 130 knots, the

17

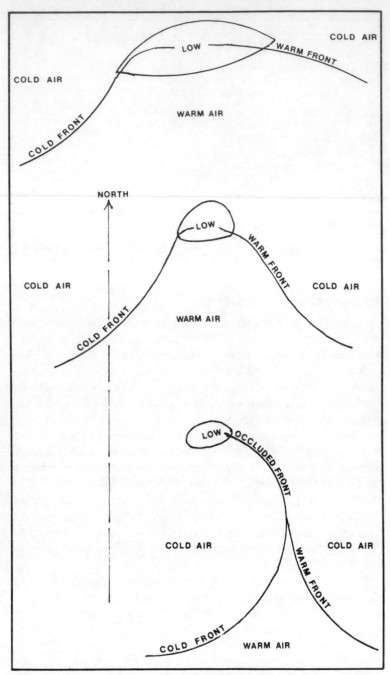

Fig. 3-3. Low pressure areas are associated with a warm and a cold front.

18

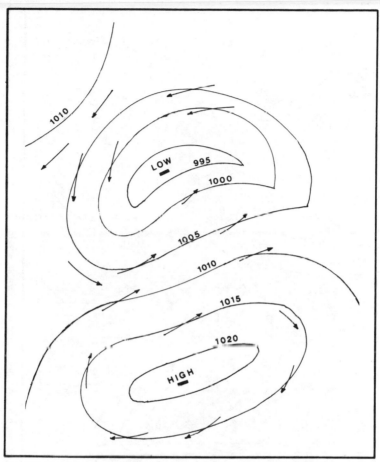

Fig. 3-4. The air circulation is always clockwise around a high-pressure area and counter-clockwise around a low-pressure area.

straight-line route would take six hours and nine minutes under no-wind conditions, but with the winds indicated on the chart, it would take something like seven hours and 40 minutes. The longer route, under no-wind conditions, would take seven hours and seven minutes. With the tailwind component shown on the chart, it would take five hours and about 45 minutes. In addition, it is likely to be a smoother and therefore more pleasant flight with the possible exception of the two times when the cold front has to be crossed.

It is a known fact that as a general rule most VFR pilots (and most IFR pilots, for that matter) don't bother with detailed studies of the weather charts which are available in all Flight Service

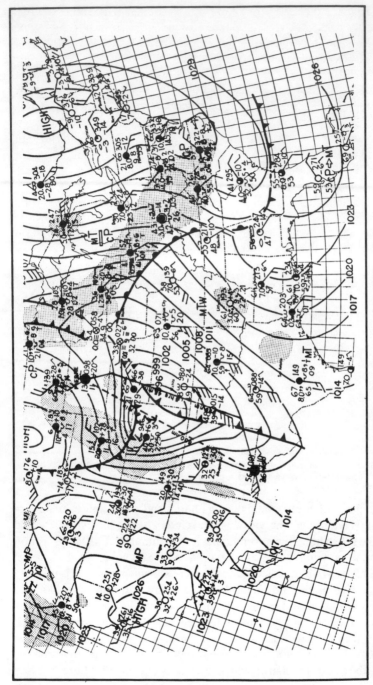

Fig. 3-5. By studying the weather charts it is possible to take advantage of favorable winds.

Stations. Practically the only time we bother with them is when the weather along our proposed route of flight is forecast to be pretty bad, and even then we are likely to depend on and accept the interpretation given by the flight service specialist. The trouble is that no matter how helpful these fellows may try to be, most of them are not pilots, and even if they are, they can't possibly know our own capabilities and the capability of the airplane being flown. Taking the lazy way out and accepting the specialist's judgment that a flight can or cannot be made VFR is no way to go about flight planning. All decisions made with reference to a flight, whether before takeoff or while in the air are, in the final analysis, the sole responsibility of the pilot in command.

In most instances, when a pilot finds himself confronted by unexpected marginal or worse weather conditions, it is because of sloppy preplanning. Granted, weather forecasts are frequently unreliable. Time and again we find that it's either much better or much worse than it is supposed to be. But by carefully examining the current conditions along our proposed route and within 50 or 100 miles to either side of that route, we can frequently devise our own estimate as to the conditions which we're likely to encounter. The more we know of the conditions all around us before starting on a flight and then during every minute of that flight, the less the chances are of running into something we can't handle safely and legally.

Chapter 4

Storms

The longer we fly, the greater are our chances of occasionally rubbing elbows with what might be referred to as severe weather. Early on in our flying careers we might make a firm decision never to fly when the weather looks somewhat doubtful. But as we pile up more and more hours we soon realize that if we want to restrict our flying activity strictly to CAVU weather, we'll be using our airplane a lot less than is necessary in order to justify our investment in flight instruction and, of course, the airplane itself.

In consequence, our concept of what constitutes flyable weather changes with time from hard VFR to more and more marginal conditions. In turn we then find ourselves confronted occasionally by severe weather and the need to know how to deal with it.

Thunderstorms

The most frequent type of severe weather, some contact with which is virtually unavoidable, is the thunderstorm (Fig. 4-1). Like most weather phenomena, thunderstorms come in a variety of shapes and sizes, but they all have one thing in common. Inside they are murderous, capable of tearing an airplane to pieces, and must therefore be avoided at all cost.

The easiest type of thunderstorm to deal with is the so-called airmass thunderstorm. It usually stands alone in solitary splendor, surrounded by clear skies and unrestricted visibility. It can be seen

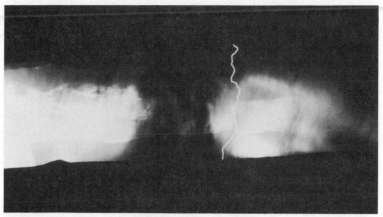

Fig. 4-1. A thunderstorm is the most frequently encountered type of severe weather.

for miles and as a result, course adjustments can be made early to give them a wide berth. Whenever possible, a detour around these storms should be planned for the side away from the direction of the anvil cloud. While this anvil, which may be as high as 40,000 feet or more, may seem perfectly harmless, there is always the chance that hail might spew from it, and hail is something which airplanes were not designed to deal with.

The worst type of thunderstorm from the point of view of the pilot is the imbedded variety because, more often than not, it cannot be seen and as a result we don't know that we're about to fly into one unless we're already right on top of it (Fig. 4-2). But since they are embedded in solid IFR weather, they should not be of consequence to the VFR pilot because there is no excuse for him to get into that kind of weather in the first place. IFR pilots, unless they are flying airplanes equipped with weather radar, have little choice but to rely on the information provided by ground-based radar and the ability of the ATC controller to vector them around the most intense areas of these storms.

The third variety encountered with any frequency is the squall line, a long line of individual storms often covering hundreds of miles in length and extending for 20 or 30 miles in width (Fig. 4-3). Occasionally there may be sufficiently wide spaces between the individual cells to permit penetration of such a squall line, but there will always be considerable turbulence and there is also always the danger of encountering hail. Most of us have flown through such openings between storms at one time or another, but it is neither

23

Fig. 4-2. Imbedded thunderstorms often cannot be seen and are therefore hard to avoid.

comfortable nor a procedure to be recommended. If we're high, there may be a place where the tops of the clouds are low enough to make it possible to stay in VFR conditions on top. This is described in greater detail in another chapter.

Fig. 4-3. A sharply defined squall line of thunderstorms can often extend for hundreds of miles.

Wind Shear

One aviation nemesis getting a lot of attention is wind shear associated with thunderstorm downbursts of air. Encountering a downburst at the wrong time under the wrong circumstances, such as low and slow during an approach, spells doom for any airplane, large and small. Remember that downbursts are in thunderstorms . . . avoid a thunderstorm and you avoid the downburst.

Research indicates downbursts are a short-lived powerful rush of air that descends perpendicular to the surface of the earth, almost always in a thunderstorm. When it encounters the ground, it spreads out creating turbulent air that wrecks havoc with an airplane.

Initially the airplane will experience a strong headwind, slowing its ground speed; then a powerful downdraft forces the airplane down; followed immediately by a strong tailwind that dramatically increases the ground speed. An airplane on approach will be low, slow, and "dirty" with landing gear, flaps, and/or speed brakes extended. An unsuspecting pilot might have a hard time dealing with the fast-changing situation and be unable to control the airplane.

Detecting downbursts within a thunderstorm is a new technology using Doppler radar that measures the velocity and direction of particles in the air, primarily raindrops. (Despite the wider safety margin offered by Doppler radar, an installation schedule will be driven by governmental economies. It will first appear at major airports and spread to others when feasible.)

Using Doppler radars, meteorologists and air traffic control personnel will be able to locate wind shear and downbursts and warn pilots of its presence. Pilots can then elect to deviate or penetrate the turbulence. Properly prepared, an airplane might safely fly through it. Or the more prudent choice would be to abandon the maneuver—like an approach to landing—and try again when the turbulence has subsided.

Tornadoes

In certain parts of the country, usually during spring and early summer, thunderstorm activity may be associated with the development of tornadoes. It should be unnecessary to say that whenever a tornado is even suspected, retreat is the only sensible course. Luckily our airplanes fly a lot faster than a tornado moves, so there is never the danger of being overrun by one. This is, of course, true of all kinds of weather. It is physically impossible for weather to catch up with an airplane. Whenever a pilot gets into weather, it is he who did the catching up, not the other way around. Actually, anyone who claims that he found himself inadvertently in weather is really stretching the truth. Granted, the weather into which he flew may have been worse than anticipated, but he certainly did know that he was flying into conditions the severity of which he may have had no way of determining in advance.

Dust Devils

Mini-tornadoes, or so-called dust devils, usually look relatively harmless but they can be quite lethal. I have seen one pick up an entire sailplane that was sitting on the ground, lift it into the air and then toss it back, breaking a lot of expensive structure in the process.

Dust devils rarely extend upward to altitudes in excess of a few thousand feet, but whenever we spot one below along our course, it is always a better idea to detour around it rather than to take a chance on having it make a grab for the airplane. And it goes without saying, when dust-devil activity is suspected, aircraft on the ground should be securely tied down.

Hurricanes

Like tornadoes, hurricanes are no place for a light aircraft. The only way to deal with them is to stay far away.

Sandstorms

One of the least attractive though not necessarily dangerous weather phenomena is the dust- or sandstorm. Such storms occur most frequently in the desert areas of the Southwest, though they do occasionally develop elsewhere. Most of the time we'll be flying along in perfectly nice VFR conditions and then we gradually become aware of a reduction in visibility and most probably, an increase in turbulence. The air around us may take on a yellowish or greyish tinge and after a while we can barely see the ground. We may be tempted to try and climb to a higher altitude in order to get on top of all that mess, but many of these storms will extend to 14,000 or 15,000 feet and occasionally even higher, making the attempt to top them fruitless. There isn't much we can do except keep on flying, primarily by our instruments, and hope that we'll be out of it after a while.

It's bad for the engine (and if we happen to be flying an aircraft with a ram-air intake, such as a Mooney, the ram-air intake should be closed). Since most such storms are generated by strong surface winds they act much like smoke, meaning that they are relatively narrow but may extend for great distances downwind from their source. In order to minimize the time for which we expose ourselves and the engine to the dust-laden air, it is a good idea to figure out the direction of the wind and then to fly at right angles to the wind rather than either into the wind or in the direction in which it is blowing. Even if such a change of direction results in a detour, it is preferable to flying in the effect of the storm for a prolonged period of time. If worse comes to worst, a landing might be the best idea. An hour or two on the ground may be all that is needed in order for the storm to dissipate.

Sometimes these storms, especially sandstorms, are clearly defined, making it relatively easy to fly around the cloud of sand and dust. But more often than not they tend to sneak up on us, and when that happens the only choice is to either relax and try to enjoy it, or to make a 180, land and wait. Dropping down to a lower altitude, such as we would normally do in order to stay below an overcast, is rarely of use. In practically all cases, these storms extend from the ground up, and conditions at low altitudes are likely to be a great deal worse than higher-up. After flying through a really bad one it would be a good idea to drop in at the nearest FBO and have him change the oil. While this won't remove all of the grit and crud, it will take care of a lot of it, and your engine will be grateful in the long run.

Dastardly Dust

Two pilots had to make a 300-mile cross country trip; departing mid-morning, stopping halfway to conduct some business, and continuing later that afternoon for home. It was supposed to be a familiarization flight for one pilot building time in a new type of airplane to satisfy insurance requirements. Luck had it that not only would he become more familiar with the airplane, but he would become familiar with flying in a dust storm as well.

A brown western sky made it obvious weather would be a factor on the trip, and they discussed who would take the left seat. The plane's owner was IFR-rated and much more familiar with the airplane, but the other pilot needed the hours. The first destination was reporting a dust storm, too, but it was still VFR, so the new pilot took the left seat with an IFR copilot.

After takeoff it was obvious the dust storm was getting worse.

Climbing through the brown air, several thousand feet high, they lost sight of the ground. Hoping to get on top of the dust cloud they climbed to 10,500, found the brown tint extended to a higher altitude and leveled off—surrounded by dark brown below and a dingy blue-brown sky above.

"Could you handle this on your own?" the copilot queried.

"No," the pilot said, shaking his head, "I would have stayed home until tomorrow."

"Well, this is a good demonstration of the value of an instrument rating," the copilot continued. "With it we can do our business and be home for supper tonight. Otherwise we would be stuck waiting for the weather to clear."

A Safe Port in Any Storm . . . .

Two airports were available for the stopover: one with a single, narrow, north-south runway, with VOR, NDB, and radar approaches; another airport with several long and wide runways and an instrument landing system (ILS). The smaller airport would be more convenient to the business meeting, but the larger airport would be safer due to size and precision approach facilities. They chose the larger airport.

The copilot called approach control and requested an IFR clearance to land. ATC complied with the request and vectored the flight around the airport . . . they were assisting a solo cross-country student pilot who had underestimated the power of the storm. As it turned out, the student landed safely and did not interrupt airport operations.

The VFR pilot flew ATC's vectors while the IFR copilot familiarized himself with the approach. Prior to initiating the approach, the copilot took the controls. As clouds of dirt moved across the airport, conditions would rapidly change from VFR to IFR. Approaching the airport, on the localizer, the pilots hoped it would be VFR so they could land on a taxiway that pointed directly into the 30-40 knot wind.

But with the approach lights in sight (it was that dark in the dust storm) the airport wen IFR, forcing a 45-degree crosswind landing on the main runway. Crabbing into the wind the copilot flew the airplane to just above the runway, pulled the power back, straightened it over the center stripe, and landed.

Homeward Bound

There was no rush to complete their business. The dust storm seemed to be letting up and every hour they waited meant better flying conditions upon departure. Sure enough, visibility improved so the second leg could be conducted VFR, but the dust was still present and a stiff west wind would be a fiendish headwind.

The second leg would be a night flight and conditions could worsen without notice so the IFR-rated pilot/plane owner climbed into the left seat to complete the journey. It proved to be a rough and slow ride with farm and city lights veiled by the dust.

Arriving at the uncontrolled home airport, the tetrahedron and windsock proved their worse fears were true. There were two runways, north-south and east-west, but only the N-S runway was lighted. Winds were straight out of the west at probably more than 30 knots. Fortunately, the plane's owner had operated from the airport for several years and knew the E-W runway's exact location.

Flying a standard rectangular pattern the pilot lined up for what appeared to be a landing in a pasture. On cue the E-W runway appeared under the landing light and he touched down without incident on the center line.

Winds subsided overnight to a morning calm, but it would take several days for the tons of sand and dirt particles to settle out of the atmosphere. The airplane's oil was changed to flush out any dirt ingested during the storm.

Chapter 5

Taking a Look

Probably more VFR weather-related accidents are the result of the pilot wanting to take a look to see if the weather conditions are better a bit farther ahead than of any other cause. It's terribly tempting. You're flying along under a 5,000-foot or so overcast, dodging intermittent rainshowers and occasional low-hanging scud. Except there isn't any turbulence, it's pretty unpleasant, but your destination a couple of hundred miles distant is reporting scattered conditions with good visibilities and you figure that sooner or later you ought to be able to get out of this mess.

Following The Interstate

There is a big Interstate highway which runs more or less along your route of flight and you stay a bit to the right of it in order to be able to comfortably keep it in sight to your left. You figure with a certain amount of justification that as long as you can see that highway for a fair distance of about a mile or so ahead, you'll be okay (Fig. 5-1).

Suddenly that group of occasionally low-hanging scud turns into something more solid and you drop down lower. For a moment that highway ahead is all gone. Then you see it again, only now you may only be a thousand feet or so above the ground. That's still not too bad but you kind of hope that it's not going to get much worse. The cars down below, especially those coming the other way,

Fig. 5-1. As long as you can see the highway below, the weather conditions cannot be too bad.

have their headlights on even though it's still several hours before sunset. That should tip you off that the weather farther on is likely to get worse than better, but it somehow doesn't occur to you. Suddenly you realize that the highway ahead just doesn't seem to be there anymore. It just disappeared from sight (Fig. 5-2). Within seconds you're right in the middle of a rainshower which makes an incredible racket as it pounds your windscreen. Straight down you can still see the cars with their headlights, but ahead there is absolutely nothing at all.

You briefly consider turning back, but figure a shower like that can't last forever. After all, it's only about another 45 minutes until

Fig. 5-2. Suddenly the highway is not there anymore and conditions are getting worse.

you get to where you want to go, and if they were right about the scattered conditions you're bound to be breaking out into sunshine pretty soon. On the average, you'll probably make it just fine. You may spend another 20 or 30 minutes being extremely uncomfortable with damp palms and dripping armpits, but by sticking religiously to the highway below and watching your airspeed indicator and altimeter more carefully than usual, you'll probably break out of this mess just before reaching your destination.

Featureless Terrains

But what if there hadn't been that highway? What if you had been flying over a featureless plain (Fig. 5-3), rolling hills or the endless woods in southern Georgia and Mississippi? The temptation to fly on and take a look would, of course, be just as strong. The reason you're in the air, after all, is that there is some place where you feel you have to be, and that some place is, as always, beyond the weather ahead. The difficulty is that you want to keep track of where you are relative to the desired course while at the same time keeping a sharp eye out for what is happening behind you just in case you find that it becomes necessary to beat a hasty retreat. Most probably there is low-hanging weather all around, but there

Fig. 5-3. But what if there is no highway to follow?

Fig. 5-4. There is something that looks like a light spot somewhere more or less in the direction in which you want to go.

is something that looks like a light spot more or less in the direction in which you want to go (Fig. 5-4). So you plow on toward that lighter area and sure enough you get to a place where the overcast seems to be somewhat thinner, but it's only a fairly small area all around the low scud and showers continue. In all probability things have closed down behind you by now and you have no choice but to fly on, staying just as low as the terrain permits. For some time now you have been unable to receive a VOR and your nav receiver is tuned to the next VOR along your route, but the OFF flag continues to stubbornly tell you that you're still beyond reception distance.

In a case like that the chances are that you'll be ending up a considerable distance off course without being fully aware of it. And trying to become reoriented strictly by pilotage is extremely difficult under those conditions. Being low, the amount of terrain features which can be seen is limited and somehow nothing seems to look the way it should look according to the chart. In addition, because you're so close to the ground, the chore of flying the airplane requires your constant attention, making the studying of charts problematical if not impossible. Whatever you do, flying the airplane must necessarily take preference over any other activity. As long as you have the airplane under control and as long as there is ample fuel in the tanks, there is always a way out. Just don't panic. Stay cool. Always remember, if all else fails and you suddenly find that there is no direction left in which you can continue on VFR, the way out (unless there happens to be a place on which you can land) is up, period.

When VFR pilots let this type of predicament develop into an eventual crash landing with often catastrophic consequences for the airplane and all aboard, it is the psychological block which prevents them from accepting the one remaining, though illegal choice, namely to climb into the overcast to an altitude at which radio reception can be reestablished. Assuming the always all-important ability to control the aircraft by reference to instruments alone, you're then in a position to contact some FSS on the ground and either own up to your predicament and let them guide you to a safe landing at an airport at which the weather is halfway decent, or you can just inquire about the current weather at the various locations within your range and then navigate by radio toward the one that sounds most likely to get you out of the soup in the shortest possible time, even if it should mean a detour or actually retracing your steps.

High Altitude Situations

Of course, not all taking-a-look situations take place close to the ground. You may be flying along at a perfectly safe and comfortable altitude, but there lying smack across your course is a squall line which extends for considerable distances in either direction. According to the weather reports, it is supposed to be about 20 miles wide when slightly to your left there is what appears to be a low saddle between the higher buildups. The temptation is to climb up and over that saddle rather than to accept the delay and extra fuel associated with flying around one end of the squall line or the other.

Most probably at that altitude you're already at full throttle, so you simply trim the nose up and start to climb. As you get closer to the clouds the likelihood exists that two things will happen at the same time. One is that there will be clouds to the right and left as well as ahead, and the second is that the saddle turns out to be higher than you thought it was. So you now steepen your angle of climb. But with clouds obscuring the horizon to either side you have now lost your ability to judge your angle of climb by visual reference to the outside. Unless your airplane is equipped with an angle-of-attack indicator, the only instrument in the cockpit that is of any value in determining if the angle of climb is getting too steep is your airspeed indicator. The VSI becomes meaningless at an altitude which may be close to the service ceiling of the airplane because the airplane is probably mushing along, showing a rate of climb of only a few hundred feet or less while it is actually flying in an extreme nose-high attitude. The artificial horizon does show the nose-high flight condition, but it fails to clearly relate it to any reference to the approaching stall. Only the airspeed indicator does that and when you find that your IAS is getting too low put the nose down, clouds or no clouds.

Squall Lines

A squall line invariably is made up of thunderstorms and if you should actually fly into the clouds, your ability to avoid the thunderstorms is reduced to a dangerous level. Therefore, if you can't get up high enough to get over the top of that saddle, turn away from the clouds. If you're still climbing at least 100 fpm or better, you might fly one or several shallow 360s (shallow because a steep bank will reduce your ability to climb) until you're certain that you can see across the saddle and can also see what's beyond. Since without visual reference to the horizon you have no certain way of knowing whether you are looking up, down, or straight ahead, it is a good idea to level the airplane by reference to the artificial horizon. Then, when it's in a level attitude, point the nose toward the saddle and look straight ahead. If the lowest point of that saddle is above the nose of the airplane, you're still too low. If it's level with the nose it would still be a good idea to climb another few hundred feet. If it's below the nose of the airplane, what do you see beyond? If the clouds beyond are clearly at the level of the saddle or lower, it's probably all right to fly on over it and continue on. If there are some individual higher buildups with big spaces in between, it may still be all right. But unless you can see with reasonable certainty that you'll be able to get through and continue on for the entire width of that squall line, you'd be a lot better advised to turn back and forget about the whole thing.

Always remember that you're probably already as high as your airplane is going to climb, so getting over still-higher cloud tops somewhere ahead may no longer be possible. Also, you're likely to be at oxygen altitudes without oxygen in the airplane and while you may feel perfectly fine and capable of efficiently handling the controls, this may in fact not be the case. Once you have taken the step and flown past that saddle, you're committed to staying up at that altitude until you're clear of the squall line. With headwinds at those altitudes often reaching 40 or 50 knots, this could possibly take quite a while.

Mountainous Terrain

A similar situation often occurs in mountainous terrain. There is an overcast and many of the mountain tops are obscured by clouds. But the visibility below the clouds is pretty good. You're flying along a valley or across a high plain toward a ridge which extends across your route of flight, most of the tops of which rise into the base of the clouds. But there is a pass where there appears to be ample space to fly through while staying safely VFR. Here too you should be level at an altitude above the base of the pass before actually entering it. The reason is the same as described above: You've got to be able to see what lies beyond. Is there a valley on the other side, is it safely VFR, and once having reached the other side of that valley, will you be able to go on?

Some of this information can be obtained by studying the Sectional charts. If you know which pass you're looking at (and that is by no means always the case) you can determine from the chart what lies ahead. If the ground falls off on the far side of the pass and there is no second mountain ridge to be crossed beyond, then it's probably reasonably safe to fly through that pass without first making a special effort to see what's beyond. But if you're not sure of which pass this is, or if the chart shows that there are more high mountains ahead, the pass should be entered only with the clear advance decision to make a 180 as soon as there is an indication that continuing VFR on course may become difficult or impossible. That last-resort escape route, climbing up into the overcast and to a safe altitude, may not work in places like the Rockies however. Here many mountains are higher than the service ceiling of many light aircraft, and clouds are bad enough, but clouds full of rocks could easily spoil a whole day.

It would be foolish to say here that one should never, under any circumstances, fly into questionable weather conditions in order to take a look. It would probably be a good rule, but it just isn't realistic. Sooner or later all of us are likely to do it. But there should be two hard and fast prerequisites without which it should never be attempted. First, the pilot must be able to control his airplane by means of instruments alone, and this involves not only straight and level flight, but turns, climbs, and descents. Second, there must be plenty of fuel in the tanks. Taking a look means not knowing what is likely to come next and thus, not knowing how long one may have to be able to continue to fly. Too little fuel turns a calculated risk into a foolish chance.

Chapter 6

Special VFR

Special VFR is something that was invented in order to make it possible for aircraft to take off or land at controlled airports when the conditions are below VFR minimums. S/VFR clearances are available at the pilot's request when the visibility in the airport traffic area is at least one mile. There is no minimum ceiling requirement (Fig. 6-1). All the pilot is expected to do is to stay clear of clouds.

In practice, S/VFR works like this. The airport is IFR and the rotating airport beacon is operating. The sky may be partially obscured by haze, smoke, fog, or smog with no ceiling. Visibility is restricted to one mile or a little better. There may be broken or solid overcast with bases below 1,000 feet and visibility below the clouds ranging anywhere from one mile to unlimited. The pilot who wants to depart on a VFR flight contacts ground control and requests a special VFR clearance.

The primary difference between an S/VFR clearance and an IFR clearance is the S/VFR clearance does not include a destination or specific altitude requirements. It will usually require the pilot to take up a specific heading immediately after takeoff and to remain on that heading, staying clear of clouds, until he has reached VFR conditions above the haze or smog or until he has left the control zone which extends approximately five miles in all directions from the center of the airport. He is expected to inform the tower when he has either reached VFR conditions on top or when he has

Fig. 6-1. Special VFR has no minimum-ceiling requirement.

left the control zone, the latter usually being a more or less educated guess based on the time it will take the aircraft to cover five miles at an average ground speed.

If the pilot fails to request a special VFR clearance and simply tells the tower that he is ready for takeoff, the tower will inform him that the airport is IFR and ask him, "What are your intentions?" The tower will not suggest that a special VFR clearance is available for the asking.

If an aircraft arrives in the vicinity of a controlled airport at which the weather conditions are as described above, and if he simply tells the tower before entering the control zone that he intends to land, he will also be told that the airport is IFR and asked his intentions. It is then up to him to request a special VFR clearance. When the clearance is given it will usually call for him to fly the downwind leg and a regular pattern, staying clear of clouds. If there is a delay he may be asked to stay in VFR conditions outside the control zone or to hold above the airport at an altitude above the ceiling of the airport traffic area (3,000 feet agl) until the clearance is available. Delays are one of the primary inconveniences associated with S/VFR, because all IFR traffic takes preference and because only one S/VFR aircraft may operate within the control zone at one time. The intervals between takeoff or approach clearances include the time it takes for each aircraft to leave the control zone or to fly from outside the control zone to the airport and to land.

Obviously, S/VFR is always associated with marginal weather

conditions, and there are times when using S/VFR can take the pilot into some fairly tricky situations.

Haze

Let's take visibility haze conditions which are frequently present at the various airports around Los Angeles but also occur not infrequently in the Northeast and elsewhere. There are actually no clouds anywhere and the sun is clearly visible above. But lateral visibility is zilch, barely that one mile which makes S/VFR permissible. The pilot is cleared for takeoff and requested to inform the tower when he has reached VFR conditions above the haze, or when he has left the control zone. If this is at one of those Los Angeles airports we talked about, the haze or smog layer is likely to be only a few thousand feet thick and he'll probably break out on top before reaching the end of the control zone. Since there are no actual clouds, the standard warning phrase, "Remain clear of clouds," is meaningless. But clouds or no, he will have absolutely no forward visibility. He may be able to see the ground below all right and the sun above, but laterally he may as well be in clouds for all the good looking out does him. Aside from keeping the wings level by reference to his instruments, the primary danger during such a climbout is the tendency to stall the airplane. With no horizon or other outside reference available to help judge the angle of climb, it is only too easy to keep applying more and more back pressure. The answer is to watch the airspeed! Never mind trying to get out of the top of this muck in a hurry. If the airspeed starts to drop, get the nose down and let it build up again. An inadvertent stall, or worse yet a stall-spin, this close to the ground more often than not is fatal. Airspeed is money in the bank and that's what has got to be watched above all.

In many other parts of the country the haze layer, especially on hot summer days, may actually be two or even three miles thick. In the average light aircraft there isn't a chance of getting on top of it. In this case the pilot reports leaving the control zone and from then on he may find himself surrounded by this awful soup for a long time, possibly the entire length of his flight. Straight down he'll be able to see the ground, but at any kind of an angle the ground quickly disappears into an indistinct blur. Lateral visibility in all directions will appear to be virtually zero even though it is likely to be several miles in reality. Since there is nothing to look at, the only way to obtain any idea of the actual lateral visibility is to look at the ground. If the airplane is at 10,000 feet agl and the ground can be seen reasonably clearly, then it is safe to assume that the lateral visibility is at least 10,000 feet or two miles. It will be less later in the afternoon when the sun is at a low angle and we are flying toward it. At such times the visibility toward the sun may be half or less of what it is away from the sun. One way or the other, flying in such haze means flying by reference to the instruments.

One of the most unpleasant aspects of such haze conditions is the difficulty involved in finding the airport. If there is a VOR or NDB right on the field it's no serious problem even though we may not actually see the runway until we're practically on top of it (Fig. 6-2). If no convenient nav aid is available, we may have to stay high until we can see the airport below us. We'll then have to circle down, keeping the airport in sight all the time. If it's late afternoon and the sun is low, we should plan this descent in such a way that the airport is away from the sun. In other words, since the sun sets in the west, we'll want to descend on the west side of the airport so that we have to look toward the east and away from the sun.

Low Overcast

Now let's look at another S/VFR situation. This one normally affects only departing aircraft. The visibility is such that we can see all the way into the next county but there's a solid overcast hanging over the airport with cloud bases at, say, 700 feet. Again the airport is IFR but the S/VFR having no ceiling minimums, an S/VFR clearance is legally available. In this case we'll level off right

Fig. 6-2. In smog or haze we often don't see the runway until we are practically right on top of it.

Fig. 6-3. If there are any hills ahead we may end up with no room between the clouds and the ground.

after liftoff and stay under the overcast until we are clear of the control zone. But what now? Whether actually reported or not, if the overcast extends for a great distance, we may find that we'll have to hedgehop for mile after mile. And that's a lousy way to fly, considering all those chimneys and broadcast towers all over the place, not to mention the fact that it's illegal over so-called populated areas. And if there are hills anywhere along our course we may actually end up with no place left between the ground and the base of the clouds (Fig. 6-3).

In such a situation the options tend to dwindle in a hurry. What we should have done was turn back as soon as we realized the apparent extent of the low overcast. But it's probably too late for that now. Instead, we keep following a road, trying to stay reasonably close to the direction in which we want to fly. But it won't be too long until, with all nav aids being beyond reception distance at our level of flight, we no longer have any clear idea of where we are.

It's decision time. What next? There aren't really a great many choices left. We can land on a road, assuming this is one of the friendly states like California of Texas where they don't mind that sort of thing and usually let us take off again from the road when the weather improves. Other states, New York for instance, are likely to insist that the plane be dismantled and trucked to the nearest airport—a less than pleasant prospect. Anyway, if a landing on a road seems preferable to whatever options are available, then it's important to make sure that the road selected is straight for a sufficient distance, and that there are no telephone poles close to the side of the road or power lines or telephone lines crossing the road. Also, a little-traveled country road is preferable to a busy highway.

Barring such a precautionary landing we can either continue assuming there still remains some space between the ground and the overcast, or we can shove in the throttle and climb into the clouds in the hopes of reaching VFR conditions on top. This latter action is strictly against the rules, but once we have been stupid enough to get ourselves into this kind of situation, altitude, regardless of the fact that it's in the clouds, is still the safest way out.

As is true of virtually all of these marginal-weather conditions, we better be able to control the aircraft by reference to instruments because at this point we are likely to have no idea of how long we'll be in the clouds before breaking out on top, if we ever do break out. Some indication of the thickness of the overcast is the amount of light there was below. If it was relatively dark, we must assume that the cloud deck may extend upward some 10,000 or more feet. Assuming that we'll be able to achieve an average rate of climb of some 400 fpm, we'll then be in the soup for 25 minutes which is a long time for a non-instrument rated pilot to be flying on the gauges.

We might point out here that in situations like this an autopilot or even a simple wing leveler is a great help. With that kind of mechanical assistance all we have to do is trim the aircraft for a comfortable rate of climb and sit back and let it do the flying for us. Not only does this prevent us from giving erratic or incorrect control inputs resulting from the lack of sense of balance when deprived of any outside visual reference, it also gives us time to study the charts and to tune in whatever nav aids we believe to be within reception distance once we get up to a reasonable altitude. When we then do finally start to receive some signals from VORs we'll at least be able to determine with reasonable accuracy where we are. Once that has been accomplished, there are a number of actions which should be taken next, assuming that we're still in the clouds and it is sufficiently dark above to indicate that we may never reach VFR conditions without having to climb to oxygen altitudes. If it is starting to get light above and there are occasional breaks, then it's probably best to simply keep on climbing and to start worrying about what to do next only after we've finally leveled off in the clear above the clouds (Fig. 6-4). But if it looks as if that

Fig. 6-4. If it is getting light and there are breaks above we might as well keep on climbing.

is not going to be possible, then we have to decide right now what to do next.

The most sensible thing to do would be to contact an in-range FSS and file instruments. There is no need to tell them that we're already in the clouds. After all, they have no way of knowing what the conditions are at your position and altitude. While that would be the right and sensible thing to do, it must be assumed that having gotten ourselves into this mess, we're not necessarily that sensible.

So if we are determined to ride this out without having to deal with ATC, then two actions should be taken without delay. One is to look on the chart and see where we are in relation to the Victor airways in the area. If there is any IFR traffic in the area, it is likely to be operating on the airways. We would be a lot better off to stay as far away from the airway centerlines as possible. This doesn't mean that there may not be IFR traffic elsewhere. Such IFR traffic, operating by means of area navigation or Loran-C, might be found anywhere at all. But the chances of running into someone are at least considerably reduced. The second action is to contact some FSS and ask for the conditions at all available reporting stations within a reasonable distance of our present position. After all, we want to get out of these clouds eventually, and the sooner we know where there is an airport with ample VFR conditions, the sooner we might be able to get back to somewhere where we can see where we're going.

In this kind of a situation, ample VFR conditions mean just that. If an airport is reporting 1,000 and three it isn't good enough. Granted, that's VFR, but hoping to break out of the overcast just 1,000 feet above the ground is asking for trouble. Under such conditions it would be necessary to fly a regular instrument approach and obviously we neither know how to do that, nor do we have the appropriate approach charts in the aircraft. As a rule of thumb it would seem that 2,000 feet between the ground and the base of the overcast should be considered an absolute minimum. More is better. If there's a nearby airport with a 2,000-foot ceiling and one farther away with a 5,000-foot ceiling, it is probably better to head for the more distant one, even though that means that we'll be in the clouds that much longer. It goes without saying that if this is hilly, or worse yet, mountain country, the danger associated with dropping out of the clouds in hopes of being able to see the ground in plenty of time to control VFR toward an airport is considerably greater. Hills and mountaintops have a way of sticking up into the base of the clouds. Just because an airport located in some valley reports a 2,000- or 3,000-foot ceiling, it doesn't mean that the surrounding terrain may not be obscured.

If we've actually managed to find ourselves in clouds above such terrain it may be necessary to plan the descent extremely carefully to be sure that we're coming down over low country and not right into the side of a cliff. The best, and in some areas, the only way to accomplish this is to pick out two VORs with a radial running from one to the other right along a valley or flat portion of ground with no obstacles in between. We then start the descent when established over one of those two VORs and fly toward the other one. As soon as we reach that other VOR we make a 180 and fly back along that same radial or bearing toward the first VOR. Throughout this maneuver we keep looking down to be sure to see the first sign of the ground as soon as it appears through the base of the clouds. This is pretty tricky stuff and the best advice is to never get caught in clouds anywhere where there are hills or mountains around.

Chapter 7

Harrowing Haze

Coping with haze can only come from experience. One low-time pilot flew a lot of eastbound trips at sunset that prepared him for a most unusual flight.

Less than 150 hours were in the pilot's logbook, but his business required flying eastbound while the sun was setting in the west. More often than not a thick haze layer developed at the temperature inversion several thousand feet above the ground. When sunlight was diffused in the haze, the horizon disappeared momentarily in a gray muck. Then slowly but surely, streetlights of cities, towns and farms would poke through the muck for a better surface reference. Finally it would get dark at cruise altitude and the haze would "disappear."

It was an eerie sight that left the VFR-only pilot uncomfortable. Too often he would rely on an autopilot to keep the airplane on course, but with more flights he became more comfortable and would hand-fly the airplane referring to instruments—while also scanning for en route traffic conflicts.

Sometimes the inversion layer carried more than haze, with cumulus clouds extending above the layer in the clear air. In this case, the pilot would stay lower than the inversion altitude, ensuring cloud avoidance when it got dark and the small buildups were not easily visible. He always looked for "veiled" ground lights or black holes that might indicate rain, or some other weather, but never had to deal with it.

However, in the future he would have to deal with extremely poor visibility, in haze, at a high altitude, in an airplane with no autopilot.

A Bottle of Skim Milk

A good weekend with friends was drawing to a close and it would be time to go home on Sunday afternoon. Unfortunately an ugly squall line of thunderstorms lurked to the west and it was moving slowly east. Satellite photos from a cable TV weather report showed the line stretching for several hundred miles lying directly on top of the pilot's route. Simple enough, he would have to wait until Monday morning when the cold front would be east of the route taking all the storms with it.

Flying directly to work dictated a dawn departure. Weather was great—no wind, stars above—driving to the airport. But as the sun started to rise, it filled the skies with whiteness, a thick moisture-laden atmosphere that made it almost like a fog or one huge cloud.

Prior to departure, visibility was adequate with no cloud cover and weather reports indicated conditions would remain unchanged through the morning. In calm air, climbing on course, visibility was about five miles.

He hoped to fly out of the haze at a higher altitude and kept climbing until reaching 8,500 feet. The haze appeared to extend another 1,000-2,000 feet higher so the plane was leveled at 8,500 with only a patch of ground directly underneath—no distinct horizon in sight.

Pilots occasionally refer to flying "in the milk bottle" when it is IFR in thick clouds and all you can see is wingtips and whiteness. The pilot decided he was flying in a bottle of "skim" milk because the whiteness was just thin enough to see down.

This airplane had no autopilot, so, unlike the previous plane that could "fly itself" in poor visibility, the pilot had to keep it on course himself. But with no turbulence and an airplane that was performing perfectly, it was easy to keep it under control.

No Hint of an Airport

Luckily it was a return trip to familiar territory and the home airport. Forward visibility was nil as landmarks passed under the wings. Normally the airport would have been visible and its tetrahedron would be in sight. There was no hint of an airport just

47

a few miles away but the ADF was pointing to the airport's NDB.

The pilot was getting skittish when the city north of the airport appeared. "The airport is supposed to be right there," he thought, but "there" was a dark gray mass that extended from the surface to above the airplane. Flying in whiteness and headed for grayness was not appealing, not to mention baffling.

A check with Unicom confirmed the suspicion that there was no rain and no wind, but why couldn't the airport be seen?

All of a sudden the plane passed through a thin cloud and emerged with the airport straight ahead in clear air—under the large shadow of a stationary cloud. Everything was visible under the cloud and now everything outside the shadow was obscured where sunlight illuminated the haze. The approach and landing were uneventful, but he would never forget the "lost" airport.

A pilot unfamiliar with the area north of the airport might have missed it. Five miles either side of it would have been away from major landmarks, only the ADF would have pointed to the airport. Imagine what could have happened to a pilot with no ADF who was unfamiliar with the area.

Considering the possibility of dense haze on future flights in unknown areas, the pilot resolved to be more careful and respect it as a hindrance to safe operation of the airplane.

Chapter 8

Between Layers

There are times when, in an effort to reach VFR conditions on top, the VFR pilot can get himself into a situation which could easily prove to be beyond his ability to handle. It usually starts off quite simply. There are low scattered clouds all over the place and it seems perfectly logical to climb up above them even though there is a higher layer of broken or even overcast clouds above. Or, at the time, there may not be such a higher layer, it may only show up some miles farther along the way. At first everything is just fine. There are patches of blue between the clouds above, and much of the ground is often visible through the spaces between the lower clouds. The change is quite gradual and, at first, the pilot may not be aware of it. Both above and below, the open areas between the clouds are becoming smaller and smaller still, until eventually there is no longer any friendly blue above or a reassuring visual contact with the ground below.

The airplane is now flying between layers, but the visibility is good and there are sure to be breaks again farther on, so we might as well continue. Wrong! This is the time to pick up the microphone and start calling people on the ground to get a truly clear picture of the weather conditions ahead and to either side. This is the time to take stock of the remaining fuel on board, to figure what range we've got and to immediately head in the direction for which conditions are reported which will guarantee that we can get down through the lower layer of clouds.

Even though the two layers between which we are flying at the moment may appear to be distinctly separate, the fact that they have turned from scattered or broken to solid is a fairly certain indication that things will continue to worsen, and odds are that somewhere ahead these two layers will gradually start to merge, becoming one thick layer rather than two thin ones (Fig. 8-1). When that happens and it nearly always does, if we want to stay legally and safely VFR, we'll have no choice but to turn around and try to retrace our steps. But there is no assurance that conditions behind us have stayed the way they were. They may, but then again the clouds may have decided to build from the bottom up or the top down, leaving us no alternative but to fly into instrument conditions.

This sort of thing is bad enough in the daytime. At night it's a lot worse because we can't clearly see what's going on around us, and we may realize that we're in the clouds only when that ee-rie glow of green and red around our wingtips is proof of what's happened.

Fig. 8-1. Odds are that somewhere ahead the two layers will merge.

Call For Help

There are no meaningful statistics about VFR pilots who ended up dead after getting caught between layers, because once such an accident has happened, the investigators know only that weather was the cause, but have no way of reconstructing what took place earlier. And those pilots who managed to extricate themselves, legally or otherwise, and who thus lived to tell the tale, usually would rather not talk about it.

And this brings us to one of the less obvious causes of many weather-related fatalities. It is the psychological block which stops the pilot from calling for help when he first finds that he has gotten into a situation which may be beyond his ability to cope. All of us hate to admit that we have done something dumb. In addition, as soon as we find ourselves in a situation which is no longer strictly VFR, we also tend to worry about the consequences which may result if we call some FAA facility and admit to being unable to continue VFR even though we're not instrument rated.

The fact is that all kinds of help is always as near as the microphone. All we have to do is pick it up and call somebody and tell them our predicament. Granted, they can't fly the airplane for us. That we have to do ourselves. But what they can do is to guide us to an area where conditions are more favorable and where we might be able to get back on the ground without endangering either ourselves or others. And if there subsequently are any legal repercussions, they are certain to be considerably less serious than a bent airplane or worse.

All of this is easier to write down than to do under actual conditions. There is always that little voice in the back of our head that keeps saying, maybe it'll get better soon. So we plow on and on until the fuel gauge tells us that soon is now, and by then it's too late to expect to have anyone on the ground do us much good. It can't be stressed too often, when things start to look bad, call for help. *Don't wait.*

Maintain Control

But when we do call for help we shouldn't mentally relinquish the control of the airplane to the controller or flight service specialist. There are scores of recorded radio transmissions by VFR pilots in trouble, where the pilot keeps repeating over and over: "Tell me what to do, I don't know what to do. . ." That is ridiculous and a sure sign of panic on the part of the pilot. We are the pilot in command and no one on the ground can tell us what to do. He can suggest a heading. If he's a pilot and familiar with the type of airplane we are flying, he may suggest a change in altitude, airspeed or such. But all he can do is suggest. It is up to us to remain calm and in complete control. This too, of course, is easier said than done, but once we let panic take hold, we might just as well point the nose of the airplane at the ground and commit suicide, because that's what may happen eventually anyway.

Chapter 9

Turbulence and Wind

Turbulence has held general aviation back more than gravity, but, like inflation and taxes, turbulence is always with us. It is virtually impossible to ever expect to complete a flight from beginning to end without getting involved in some degree of turbulence. Though the average turbulence, while uncomfortable, does not represent any danger, it is important to know how to deal with it when it suddenly decides to hit us full force.

Turbulence comes in all shapes and sizes from friendly little choppiness to incredibly hard bumps to continuing series of up- and downdrafts. It tends to be most frequent and intense near the ground (Fig. 9-1), though there is clear-air turbulence reported occasionally at altitudes up to 10 and 12 miles above the surface of the earth. While we know that turbulence is the result of movement of the air, either in the form of wind, suction produced by thunderstorms, or thermal activity which develops when the sun heats the ground, the study of turbulence, especially clear-air turbulence is, at best, an inexact science.

We can't see turbulence unless it is associated with visual phenomena such as clouds, smoke or blowing sand and dust. We can form educated guesses about the areas in which turbulence will be strong when known winds blow across mountains or other ground-based obstructions. We have read about and may have experienced the wake turbulence produced by all powered aircraft and have been warned about the serious dangers associated with

Fig. 9-1. Turbulence tends to be worse close to the ground.

the wake turbulence created by slow flying heavy jets during the takeoff and landing phase. But we have no way of knowing or avoiding clear-air turbulence. It is thought to be the result of waves of moving masses of air coming up against denser air and breaking, much like ocean waves break when they approach the shore. No one seems to know for sure, and, in the final analysis, knowing what causes clear-air turbulence wouldn't be of much help in avoiding it. If we get hit by it, all we can do is slow the airplane down to maneuvering speed and ride it out. Thus, since there is nothing much we can do to avoid clear-air turbulence, and since most of that type of turbulence is experienced at the higher jet altitudes, we'll concentrate in this chapter on the kind of turbulence which affects most general aviation aircraft, that which is produced by wind or temperature variations and is found at altitudes from zero to 15,000 feet or so.

Wind, when thought of as a simple movement of air, is neither dangerous nor particularly bothersome. As a matter of fact, since movement of air, or in other words, wind over the airfoil of the wings, is what keeps us aloft, flying without such so-called relative wind would not be possible. What causes wind to take on the character of an adversary is the interaction of this movement of air with other forces and facts.

Let's take these two forces, wind and turbulence, and examine them and their effect on our activity as pilots, from the ground up.

Taxi

Winds of average velocities have little if any effect on the aircraft when it is taxiing on the ground. Stronger winds, say, from 15 or 20 knots on up, will make it necessary to hang onto the yoke to prevent the rudder, ailerons and, when taxiing downwind, the elevators, from being slammed hard against their stops. When taxiing in a strong crosswind it is advisable to turn the aileron control as far as possible into the wind in order to reduce the lift which is being produced by the upwind wing. Under really strong wind conditions it may be necessary to have someone walk along and hang onto the upwind wing, especially when the aircraft involved is a light high-wing single, such as a Cessna 150 or 152, and this is even more important in gusty conditions. As a general rule, heavier aircraft and low-wing airplanes are easier to taxi than are the light high-wing ones.

Runup

During the runup prior to takeoff the airplane should always be faced into the wind so that sudden gusts don't have a chance to grab hold of the control surfaces while we are busy with the various pre-takeoff checks. On warm days this also helps to prevent excessive heat from accumulating in the engine compartment while we are waiting for the takeoff clearance.

Takeoff

The takeoff figures published in the owner's manual of all aircraft are based on certain assumed conditions, regardless of whether they are referring to the ground run or the distance required to clear a 50-foot obstacle. Unless otherwise specified, these conditions assume that the aircraft is at gross weight; that the runway is paved, smooth and dry; that the takeoff is being made at sea level with an outside air temperature of 59 degrees F (15 degrees C); and that there is no wind. They also usually state the recommended amount of flaps to be used.

Each variation in any of these assumptions will affect the take-off distance. If the aircraft is below gross, the takeoff run will be reduced, though probably not very much. If the runway surface is rough or wet, or if the runway rises by a few feet from the takeoff end to the other, more distance will be required. Similarly, if the temperature is above standard, it necessitates a longer ground run to achieve liftoff speed, and the same holds true when the airport is at a higher-than-sea level elevation. But by far the greatest variable is wind. Every knot of available headwind will increase the speed of the relative wind over the wings and, in turn, reduce the takeoff distance. This remains true as long as the wind is directly on the nose of the aircraft or, at least, not more than 30 degrees or so to either side. A strong crosswind, even though it may be blowing at less than a 90-degree angle, tends to have the opposite effect because the control-surface deflections needed in order to keep the airplane going straight down the runway create a certain amount of drag which increases the time and distance necessary for the airplane to accelerate to flying speed.

Airplanes have a published maximum crosswind component beyond which it is not recommended that takeoffs or landings be attempted. This, too, is based on the assumption that the aircraft is at gross and that the runway surface is dry. While expert flying technique may make it possible to successfully take off or land under somewhat higher crosswind conditions, less weight in the aircraft or slippery runway surfaces may call for a reduction in the maximum crosswind component. Under such conditions the width of the runway should also be taken into consideration. If the runway is 150 or 200 feet wide, the pilot of a single-engine aircraft may be able to start his takeoff run on the downwind side of the runway and angle his airplane several degrees more into the wind than would be possible by aligning it with the centerline. But this works only if the runway is dry and smooth. If it is slippery because it is wet, covered with even slight amounts of snow or ice, or bumpy in a way which would cause the wheels to intermittently lose firm contact with the ground, it may actually be better to start the take-off run on the upwind edge of the runway, pointing the nose along that edge of the runway, and accepting a certain amount of drift across the width of the runway as the result of the crosswind.

In addition to the above, under strong crosswind conditions, very little or no flaps should be used. Especially when the wind is gusty, flaps tend to create too much lift too soon. As the result of a gust, we may find ourselves lifting off momentarily before the aircraft is actually ready to fly. Try to hold the airplane firmly on the ground with all three wheels until slightly higher than normal liftoff speed has been achieved. If the nosewheel is lifted too soon, the airplane may try to weathercock into the wind, making control even more difficult.

As soon as the airplane is airborne, assuming there are no major upwind obstacles, turn the nose into the wind. Do not attempt a downwind turn until ample flying speed and altitude has been achieved. The reason for this requires a bit of explanation.

Even though, once airborne, the aircraft theoretically moves with relation to the block of air within which it is located and not with relation to the ground, the ground, in this case represented by gravity, does exert a degree of influence on the movement of the airplane. Thus, while the airplane theoretically continues to fly at a steady airspeed regardless of whether it is turning upwind or downwind, it does take a moment for the airspeed in the new direction to stabilize. For instance, if we are climbing out at 100 knots with a 25-knot 90-degree crosswind, when we then make a 90-degree turn into the wind the airspeed will momentarily increase—not all the way to 125 knots, but to some speed above 100 knots. The airplane, as a result, will want to climb more steeply. Conversely, if we make a 90-degree turn downwind, we'll momentarily lose some portion of that 25 knots and the airplane will tend to sink, with that tendency being further aggravated by the bank angle of the wings. Thus, such a downwind turn after lift-off should not be attempted until we are high enough to accept a degree of altitude loss without endangering ourselves or the aircraft.

Climbout

A similar effect can be experienced during the early phases of the climbout if the wind is disturbed by ground obstacles such as hangars, trees, hills, etc. It then tends to burble with constant and unpredictable variations in direction and velocity, more often than not having a detrimental effect on the airspeed and, in turn, the ability of the airplane to gain or maintain altitude.

The lapse of time in airspeed adjustment to a change in wind direction, whether as the result of a turn executed by the pilot or of a so-called wind shear, varies to some degree with aircraft weight. Heavy aircraft, such as airliners, take longer to regain their original airspeed than do light aircraft, and for them the effects of low-level wind shear are therefore likely to be more serious.

Wind shear is the dividing line between two air masses moving in different directions. This is experienced most frequently close to the ground, where the ground wind may be blowing from, say, southeast while the wind only a few hundred feet above the surface may be from the southwest. Even complete 180-degree differences in wind direction are not at all unusual. While such wind shear, when experienced at a reasonable altitude, is nothing much to worry about other than that it may bounce us around a bit, close to the ground it can be critical. Since it is rarely predictable, the pilot should always prepare for wind shear by obtaining and maintaining an ample margin of extra airspeed.

During climbout we gradually move from the area where the winds and the associated turbulence are directly affected by ground-based obstacles into the realm where the winds are usually stronger but where the air, more often than not, is smoother. Whether we prefer to climb steeply at, say, the best rate of climb, or more gradually at a cruise-climb is largely a matter of personal preference, though it may still be affected by certain other factors. Since the visibility over the nose of most aircraft is less than satisfactory when the aircraft is in a nose-high attitude, we may prefer to keep the nose somewhat lower, especially if we are climbing out in an area where there is much traffic. Conversely, on hot days when low-level thermals can cause extremely uncomfortable turbulence within several thousand feet of the terrain we may want to get up as fast as possible and may, therefore, prefer to use a steeper angle of climb.

Though psychologically we seem to feel that cruise climbing is the more economical method of getting to our cruising altitude—because while climbing, we are also covering a lot more ground—this is actually a misconception. The difference in the overall time en route as well as the amount of fuel consumed between a steep and a shallow climb is so minimal as to be meaningless in most aircraft. Thus, the angle of climb is a matter of pilot preference rather than economics or time.

Cruise

The cruise portion of the flight is usually the longest in terms of time and the portion during which most of the fuel is being burned. It is, therefore, also the portion of the flight during which we want to be as comfortable as possible and take the greatest advantage of favorable winds or try to minimize the effects of unfavorable winds.

Comfort means finding a flight level at which we can expect to encounter a minimum of turbulence. Unless there are clouds to tell us what's happening with the airflow aloft, this is not always easy. On hot summer days when those puffy clouds are spread like polka dots all over the otherwise clear sky, it is usually extremely uncomfortable below those clouds, but quite smooth above. On such days, even if there is insufficient humidity in the air to produce clouds, we can still expect a turbulent layer from 5,000 to 10,000 feet, with smoother and more comfortable conditions at higher altitudes. This does not always work in the mountains where thermal conditions often reach much greater heights while the practical limits of our cruising altitudes puts us near to the higher terrain (Fig. 9-2). The fact is that flying the Rockies on a hot summer afternoon is something only a masochist would enjoy.

There are other conditions, often associated with some type of frontal activity, when the air near the ground may be smoother and more comfortable than at higher altitudes. What is happening then is that high-level variations in temperature, which are the causes of whatever frontal activity may be taking place, generate winds of differing direction and velocities, causing turbulence which may be visible in the form of cloud patterns or may be invisible in clear air.

Much wind-related turbulence is concentrated in areas of hilly or mountainous country. Here the surface winds are pushed upward

Fig. 9-2. Flying over mountains on a hot summer afternoon can prove to be most uncomfortable.

as they hit the hills or mountains, in turn pushing the air above farther upward. Once past the peak, the air tumbles down much like a breaking ocean wave, pulling the air above it down in turn. In consequence, even though we may actually be cruising several thousand feet above the highest portion of the terrain, we are still likely to be subject to ground-induced turbulence.

Among and near the higher mountains we are faced with unique wind and turbulence problems. Winds blowing through passes and valleys are subject to a venturi effect which can, at times, increase their velocity by as much as 100 percent. Winds blowing across ridges will cause sustained and often smooth updrafts pushing us upward at 1,000 or more fpm. But as soon as they pass the top of the ridge, they'll spill down just as steeply, often resulting in downdrafts which may exceed the climb-capability of the aircraft. In this type of country, trying to escape the turbulence, whether resulting from heat or wind, is a hopeless endeavor unless we're flying a jet or, at least, a turbocharged aircraft equipped with oxygen.

One unique and potentially dangerous phenomenon associated with mountains is the so-called mountain wave. Soaring pilots seek out mountain waves because of the tremendous altitudes which can be reached by riding them. What happens is that fast moving air hits the side of a mountain and flows upward in continuous waves which may extend for great distances beyond the actual mountains. The upper portions of such mountain waves are generally smooth and may result in a continuous updraft which may require a throttle reduction if we want to stay out of oxygen altitudes. But below the mountain wave on the lee side of the mountain, something quite violent develops. Here we find an area of turbulence, referred to as the rotor, where the fast-moving wave air conflicts with the normally undisturbed air at lower levels, causing rolling and tumbling air masses which, in extreme cases, may actually be violent enough to tear a light aircraft apart. The danger signs are lenticular clouds. These distinctively shaped clouds—they look like lenses, hence the name lenticular—tend to sit right on top of these areas of the worst turbulence. They appear deceivingly calm, but the prudent pilot should carefully avoid flying under them.

While mountain waves of serious proportions are usually only found over the higher ranges such as the Sierra Nevada or the Rockies west of Colorado Springs, the wave effect can, to a minor degree, be experienced in the vicinity of all kinds of smaller mountains and even hills.

So far we have talked primarily about turbulence during the cruise portion of the flight, and how to avoid the worst of it. But wind alone, even when not associated with any appreciable turbulence, is something which must be reckoned with.

As a general rule, weather moves across the United States in a semi-circle from the northern Pacific southeastward toward the Gulf of Mexico and then turns northeastward toward New England before disappearing over the Atlantic Ocean. This means that the prevailing winds, at altitude, tend to be northwesterly in the western portions of the country, westerly in the central and southcentral states, and southwesterly in the east. This flow, with occasional variations can be expected to exist at altitudes of 18,000 feet and above. Most of the time the velocity of these winds is strongest between 35,000 and 45,000 feet, decreasing both above and below (which is the primary reason why Learjet and others have spent large amounts of money to obtain certification to altitudes up to 51,000 feet).

For all practical purposes at the altitudes at which most of us like to fly, we can expect to fight headwinds when we're heading west, and tailwinds when going east. Thus it frequently makes sense to plan on higher cruising altitudes for easterly flight, but lower levels when going the other way. But, as all of us manage to figure out only too soon, there are never enough tailwinds to make up for the headwinds. One reason is that even if we fly identical distances in opposite directions, one with a tailwind and the other with an identical headwind, the headwind always wins out. Figure it this way: We're flying a distance of 300 nm in an airplane cruising at 150 knots. With a 20-knot tailwind we will cover that distance in one and three quarter hours. But when turning around and flying back with a 20-knot headwind, it'll take us two hours and 18 minutes. This means that we'll be exposed to the detrimental effect of the headwind for 33 minutes longer than to the advantageous effect of the tailwind.

In practice, especially with the high cost of today's fuel, this translates into certain sensible operating practices. When able to take advantage of a tailwind the sensible thing to do is to slow down to a speed which approximates the ground speed which we could have expected in non-wind conditions, thus giving ourselves the opportunity to take advantage of the free push for a longer period of time. When fighting a headwind, on the other hand, it is prudent to increase the airspeed despite the attendant increase in fuel flow, because this will decrease the time during which we are being held back by the flow of the air.

Wind components also determine, to some extent, the altitudes which we should select for cruise. A light headwind at altitude may be preferable to no wind or even a light tailwind close to the ground, because we burn less fuel with the lean mixtures at higher levels of flight, and at the same time achieve a meaningful increase in true airspeed, probably sufficient to negate the wind differential. But if the wind high up is considerable, say 15 knots or more, we'd probably fly faster and cheaper (though not necessarily more comfortably) at a low altitude, despite the higher fuel consumption at the necessarily richer mixtures.

Aircraft equipped with turbochargers and supplementary oxygen can frequently, especially on eastbound flights, take advantage of some pretty sensible winds. On one flight from Portland, Oregon, to the East Coast in a turbo Centurion, I remember moving across the ground at a speed in excess of 450 knots. The airplane has a normal TAS of somewhere between 155 and 160 knots. I was flying at 23,000 feet where the tailwind component apparently added up to close to 300 knots. The flight was smooth as silk and it was amusing to have controllers on the ground asking. "What kind of an airplane is that you're flying?" But these days, with all airspace above 18,000 feet being under positive control, such flying does require an instrument ticket and appropriate instrumentation in the airplane.

Since winds-aloft forecasts are notoriously inaccurate, keeping track of the actual wind and its effect on ground speed is especially important during extended cross-country flights in the slower type of aircraft with relatively limited range. To illustrate, a 100-knot airplane with a no-reserve range of four hours, can theoretically cover 400 nm. By sticking to a prudent 45-minute reserve, this range is reduced to 325 nm. If we now add a wind factor of, say, 25 knots on the nose, the remaining ground speed is 75 knots which, with 3.25 hours of flying time, results in a range of only 244 nm. By contrast, a 130-knot airplane with sufficient fuel for six hours (or 5.25 hours with 45-minute reserve), can cover 682.5 nm in no-wind conditions and 551.25 with a 25-knot headwind. In the first instance, the range reduction amounts to 25 percent. In the second example, the headwind cuts the range by somewhat less than 20 percent. This shows that, in terms of percentages, the ground speed and, in turn, the range of an aircraft is affected to an increasingly lesser degree the higher the available airspeed. The tendency of many pilots to preplan refueling stops and then to be loath to change those plans is a bad one. We always have to remain sufficiently flexible to change our minds when wind conditions are other than had been expected or forecast.

Descent, Approach and Landing

The final phase of any flight is always thought of as the one requiring the greatest concentration, and here is where winds can play unpleasant tricks. In order to be prepared, to know what to expect once we embark on the final stages of the approach, it is helpful to know in advance the surface-wind conditions at the airport. If it is a controlled airport, the tower will provide this information once we announce our intention to land. If it's an uncontrolled field the Unicom operator may do the same thing, but it must be remembered that his, most probably, is simply a guess and may, therefore, be only an approximation. Where no Unicom service is being provided the best we can do is call the nearest FSS and use the information provided by them, remembering that, depending on the distance of the FSS from the airport, their information may not be accurate.

What it's all about is the effect which shifting or gusting winds have on the pattern and the final approach. Assume that Runway 36 is the active runway and the wind is blowing from the northwest at 20 knots. Depending on the speed of the aircraft and the exact direction of the wind, we may find that we have to hold a 20-degree correction to the right while flying the downwind leg. This means that the change of direction from downwind to final approach is not the usual 180 degrees, but rather 220 degrees, because on final it will require the same 20-degree correction, except this time it is to the left. Along the way the base leg is likely to give us trouble. If we have flown the downwind leg at the usual distance from the runway, the wind may cause us to drift past the final-approach path in which case we may have to turn even further in order to get lined up with the runway. In this kind of a situation it is often advisable to fly the downwind leg somewhat farther from the runway than usual and we may find that we want to skip the base leg altogether and to substitute a continuous relatively shallow turn.

Keeping the turns shallow, even if that might mean that we end up having to go around and try again, is important because if we see that we are about to drift past the final-approach path we are likely to want to steepen the bank. That impulse, and the usually associated impulse to use bottom rudder, should be resisted because there is always the ever-present danger of getting into a stall-spin, recovery from which would be impossible at this altitude. Also remember that frequently there is a difference in wind velocity at different distances from the ground. If the wind velocity increases rapidly with altitude, as is frequently the case, the upper wing of an aircraft in a steep bank may actually be surrounded by air which is moving faster than the air affecting the lower wing. This would increase the lift produced by that upper wing and cause the bank to steepen even further.

The best practice during gusty crosswind conditions is to fly a wider and longer than usual pattern in order to give sufficient time to make whatever adjustments are necessary without haste. Under these circumstances a make-short-approach clearance by the tower should always be rejected.

While it is generally preferable to fly a complete pattern, few of us would be inclined to reject a straight-in approach when it is given by the tower. The advantage of such a straight-in approach is that we don't have to bother with those tricky turns into or away from the wind. The disadvantage is that we may set up what appears to be just the right rate of descent to touch down on the numbers only to find that, as the result of changes in wind velocity, we suddenly sink at a faster rate than expected and are in danger of undershooting. The exact opposite can also occur. During such a long straight-in approach, or, during any final approach, we must always be prepared to make instant corrections (Fig. 9-3).

Again, less than usual flap settings, or no flaps at all will simplify the task of landing in gusty or crosswind conditions. Full

Fig. 9-3. During every approach when we get close to the surface we must always be prepared to make instant corrections in the event of sudden wind shear.

flaps should be avoided if at all possible because a sudden gust hitting those flaps, especially those huge barndoors on the lighter Cessnas for instance, could make control of the airplane extremely difficult. And, during the final phase of the approach, we should never be tempted to lower the nose. Hold the nose up and, if excessive sink is being experienced, add a bit of power. Lowering the nose could easily result in a touchdown of the nosewheel while the mains are still in the air. At best the airplane will then start to porpoise or wheelbarrow with the pilot frantically trying to use the controls to overcome the situation. At the same time there is virtually no control over the direction in which the airplane is moving, because all three wheels must be firmly in contact with the ground in order to provide that directional control. In addition, with nosewheel structures being much less sturdy than those of the mains, a hard nose-low touchdown could easily break the nosewheel and, in turn, damage the prop, making the whole affair a rather expensive arrival.

Wind in a million varieties is the primary reason why no two landings are ever exactly alike. The best we can do is to be prepared for the unexpected and thus minimize the chances for unpleasant surprises.

Chapter 10

Heat and Cold

Temperature is an important factor in aviation. It affects the way our airplanes behave during virtually every phase of flight. Temperature creates changes in the composition of the air, and without air airplanes can't fly and engines won't run.

Normal atmospheric conditions, those used to determine all those performance figures listed in the owner's manuals, are 59 degrees F (15 degrees C) and sea-level altitude. Any variation from this norm will cause variations in the manner in which the airplane flies and the engine performs. The major components of an airplane which react to air, which need air in order to perform their functions, are the airfoils and the engine (Fig. 10-1). The airfoils are the wings, the horizontal and vertical stabilizers, and the propeller blades. Air flowing over the wings creates lift. Air flowing over the horizontal and vertical stabilizers creates directional stability in both the horizontal and vertical plane. Air moving past the rotating propeller blades results in thrust.

The engine must be fed a mixture of fuel and air in an approximate proportion of 15 pounds of air to each pound of fuel. The less air available, the less power the engine is able to produce.

Density Altitude

When the air temperature is higher than normal, the air is thinner. The number of molecules contained in a given unit of air

66

Fig. 10-1. Major components of an aircraft that need airflow to perform their major functions are the airfoils and the engine.

is reduced and, in turn, the effect of the air on airfoils and engines is reduced. We refer to this condition as density altitude because air loses density not only with increasing temperature but also with increasing altitude. Thus, when the temperature at sea level is around 100 degrees F the conditions are equal to those at 3,000 feet altitude or, to put it another way, a sea-level airport at that temperature has a density altitude of 3,000 feet. And at 3,000 feet the maximum performance capability of the average normally aspirated engine and the airfoils is reduced by about 20 percent. In practice this means that an airplane with a takeoff distance of 1,000 feet will require 1,200 feet and the rate of climb, immediately after liftoff, will be 20 percent less.

One might argue that thinner air would reduce the adverse effect of induced and parasite drag. While this is true, the improvement in performance due to lessened drag is so minimal as to not be worth serious consideration.

The Effect of Temperature

Let's take the various phases of flight and analyze the effect of temperature. It is probably most serious during takeoff, especially when the takeoff is to be made from a high-altitude airport. The greater the temperature, the longer the airplane will take to accelerate and once airborne it may not achieve a decent rate of climb for an inordinately long time. On a hot summer afternoon at a place like Denver, with its 5,000-foot-plus elevation, the density altitude may be 8,000 feet or more which is only a few thousand feet below the service ceiling of many of the lower-powered, single-engine aircraft. The term service ceiling is the altitude that an airplane, under normal atmospheric conditions, will achieve a rate of climb of 100 fpm. Obviously, then, we can't expect much of a rate of climb if we start our flight close to the service ceiling in terms of density altitude. The wings don't have enough airflow to produce adequate lift and the manifold pressure, representing the air pressure needed by the engine in order to produce power, is too low to achieve normal operation.

Two facts are of major importance under these circumstances. One involves understanding the readings produced by the airspeed indicator. The relationship of the indicated airspeed readings (IAS) to liftoff speed, stall speed, etc., does not change with density altitude. If an airplane lifts off under normal altitude and temperature conditions at, say, 65 knots IAS, it will lift off at that same 65 knots IAS on a hot summer day at a high-altitude airport. But, even though the IAS is the same, the true airspeed (TAS) and the ground speed will, in fact, be much higher. A pilot who is not accustomed to operating under these conditions should ignore the visual impression resulting from the great speed with which he is moving across the ground, and be guided solely by the airspeed indicator to determine when he has reached sufficient IAS to facilitate liftoff. Conversions from IAS to TAS should only be made after a safe cruising altitude has been reached.

The other important consideration has to do with the engine. Most of us who normally operate from airports located at fairly low elevations will automatically tend to want to operate the engine at its full-rich mixture during the takeoff phase of the flight. But if the density altitude is high, say about 5,000 feet or more, then this mixture will be much too rich to permit the engine to function properly. At best it will produce less power than would be available with a leaner mixture. At worst it can result in spark-plug fouling within minutes which can cause the engine to sputter or actually fail at a time when such an engine malfunction might easily have catastrophic consequences. Therefore the pilot should lean the mixture during the pre-takeoff runup. In aircraft equipped with exhaust-gas-temperature gauges, a good rule of thumb would be to run the engine up to full throttle on the ground and lean until the gauge reads approximately 100 degrees on the rich side of peak. And he should then use that mixture setting for his takeoff run. In aircraft without an EGT gauge and with fixed-pitch propellers, the mixture should be adjusted until the highest rpm is achieved with the throttle full in. In non-EGT-equipped aircraft with constant-speed props, the only available indication would be the sound of the engine.

On hot days like that, especially when there is little wind, prolonged operation of the engine on the ground can cause it to overheat rapidly. If a cylinder-head-temperature gauge is available, be sure that its indication remains in the green. Too much heat in the engine compartment is not only detrimental to the health of the engine, it also reduces the amount of power being produced. On airplanes without a cylinder-head-temperature gauge, the oil temperature, while less reliable, does give an indication of engine temperature.

During the cruise portion of a flight, the air temperature is of less concern than during the takeoff or landing phase. If it's hotter than normal at altitude, the service ceiling and the absolute ceiling (the altitude at which the aircraft can barely maintain level flight at full throttle) will be somewhat lower than at colder temperatures. In flat country this will be of no great concern. In the mountains on the other hand, it may mean the difference between being able to get over some high terrain or having to fly around it.

Temperature also has an effect on the readings which we get from our altimeter. When it's very cold, the altimeter will read higher than we actually are. This error is not very great, some four percent for each 20 degrees F, but it can be critical when, on a cold day, we judge our ability to get over an obstacle by the reading obtained from the altimeter. In practice, if we're flying at 10,000 feet according to the altimeter, and the outside air temperature is minus 20 degrees F, our actual altitude will be 9,000 feet. A simple way to remember this is: The lower the temperature, the lower we are in relation to what the altimeter is telling us.

How To Gauge Landings

When coming in for a landing, the effect of high density altitude is similar to that at takeoff, though the reasons are somewhat different. While the takeoff requires a longer-than-normal distance because of the reduced power available from the engine, this power availability is of no consequence when landing. But, because of the vastly increased difference between IAS and TAS, we'll actually be touching down at a much higher TAS and ground speed, thus needing a lot more runway in order to slow down the airplane.

Again, in setting up our final-approach speed we should take care to ignore the speed indication received by visual cues from the outside and use the IAS which is the right one for the approach and which remains unaffected by density altitude. (Technically, though, this is incorrect. The IAS is affected by the density of the air. It simply needs a greater TAS in order to produce the comparable IAS. Therefore, from the point of view of the pilot, no increase in IAS is needed during high density-altitude conditions in order to achieve the required liftoff or approach speed).

During the summer, when high-temperature days are the rule rather than the exception, it is advisable to plan most flying activity for the early morning hours before the heat of the day. Not only will the airplane perform better, it is also a lot more comfortable because heat-related turbulence is at a minimum. And an additional advantage is that most thunderstorm activity is generated during the afternoon and early evening.

Chapter 11

Wake Turbulence

There has been so much talk about wake turbulence that it seems superfluous to spend much time on the subject here. Still, it does constitute a condition of potential danger, so let's briefly look at the best way to avoid it.

Wake turbulence is a condition which results when large heavy aircraft, predominantly jets, are flying in the slow takeoff or approach and landing configuration. It consists of two horizontal tornadoes, generated by the wing tips of these aircraft, and rotating in opposite directions toward one another (Fig. 11-1). In still air they may remain behind the aircraft and slightly below its flight path for several minutes. Wake turbulence must not be confused with jet blast. It is generated only when there is weight on the wings, meaning that there is no wake turbulence being generated as long as the wheels of the aircraft remain in contact with the runway, or as soon as the wheels touch down during a landing.

Landing Behind a Heavy Jet

With this in mind, a light aircraft operating behind a heavy jet should adjust its flight path to stay out of the areas in which these invisible horizontal tornadoes can expect to be found.

When cleared to land behind a jet, the tower will usually include a warning "caution; wake turbulence from landing 747" as part of the clearance. The routine to follow is this:

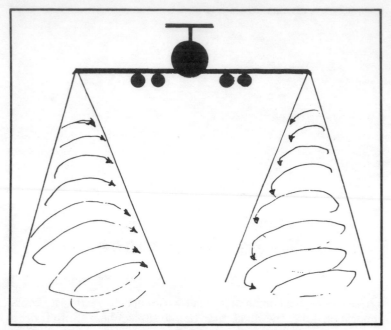

Fig. 11-1. Wake turbulence is produced by slow-flying heavy jets.

Stay above the flight path of the landing jet (Fig. 11-2). Note the point on the airport at which its wheels touch down. Then plan your own landing in such a way that you touch down beyond the point at which the jet's wheels touched down. This may be a fair distance down the runway, but there should be ample room for you to bring your aircraft to a stop, considering that the jet, too, must have sufficient runway left after touchdown to slow to a speed which permits it to turn off the active. If you feel funny about staying fairly high above roughly half the runway, you might want to tell the tower that you plan to land long, beyond the jet's touchdown point.

WAKE ENDS

Fig. 11-2. When landing stay above the flight path of the jet.

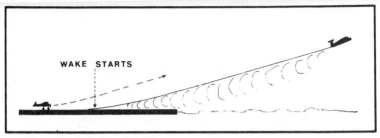

WAKE STARTS

Fig. 11-3. When taking off, lift off before the point where the jet's wheels left the runway.

If there is a crosswind, the wake turbulence will move downwind of the actual approach and landing path. In that case, staying slightly upwind of the flight path of the jet will add a degree of extra safety.

Takeoff After a Jet

When given takeoff clearance behind a departing jet, the tower will usually include a similar warning in the clearance. In this case the routine would call for you to lift off before reaching the point at which the jet's wheels leave the runway. Then climb steeply to stay above the jet's departure path, and, as soon as possible, turn to one side or the other, away from that path. If there is a crosswind, always turn into the wind to put as much distance as possible between yourself and the departing jet's wake turbulence (Fig. 11-3).

All aircraft, no matter their size, create a certain amount of wake turbulence, but it's only serious in the case of large jets. Once these aircraft have cleaned up all those high-lift devices they use during the landing and takeoff phases, the amount of wake turbulence generated and is reduced considerably and is no longer considered dangerous. But you may still experience a sudden and quite violent bump if you should accidentally stray into a portion of air through which a jet flew a few moments earlier. Therefore, when operating in the vicinity of airports with jet traffic, it is always a good precaution to keep track of where jets are and where they are going, and to give them a wide berth.

Chapter 12

Flying the Mountains

While there is nothing marginal about mountain country for the pilots who must deal with mountains all the time, for eastern and midwestern pilots who do most of their flying in parts of the country where altitudes below 5,000 feet will clear all obstacles, a first flight across the Rocky Mountains is quite an eye-opener.

There is something both scary and exciting about looking at charts which are printed mostly in varying shades of brown. One's mind's eye can't help but visualize towering peaks, steep cliffs and deep valleys, which shrink the size of the airplane by comparison.

Questions come to mind which have never been important, as long as one stayed in the flatlands. Will the service and the absolute ceiling of the airplane clear the highest point along the planned route? How long will it take to get up to those altitudes, and how much runway will be needed to get safely off the ground when the airport is at 6,000 or 7,000 feet msl? And what about the weather? It is likely to be much like that in the rest of the country, or is it, too, different? And the winds, what about those up and down drafts that people keep writing about?

These days, with more and more business and industry concentrated in the west, the need to negotiate this high-altitude third of the country is becoming more and more important to an increasing number of pilots. That many attempt these flights while being inadequately prepared is borne out with depressing frequency by reports of light aircraft, many of them high-performance singles and light twins, which are missing and often not found for months.

Still, mountain flying can be perfectly safe if adequate precautions are taken and if the pilot bothers to acquaint himself in advance with a few special techniques.

Preplanning

For the purposes of this chapter, let's plan a flight from Omaha to San Diego in a Cessna Skyhawk. This covers much rough country and the aircraft, though marginal for mountain flying, continues to be the most popular and numerous in the general aviation fleet.

First, we need charts with the most detailed topographical information—in other words, Sectionals. WACs may be acceptable for the more experienced, but only pilots who know the Rockies like the back of their hand should attempt such flights equipped with the low-altitude radio-facility charts (and even for them that is not the most intelligent thing to do).

To permit intelligent planning of the route to be flown, first analyze the performance parameters of the aircraft. The Skyhawk, vintage 1975, has a maximum no-reserve range of 773 nm at 55 percent of power. At this power setting it produces a TAS of 103 knots. Prudence dictates making a fuel stop with at least one hour reserve left in the tanks. Furthermore, we know from experience that there will be headwinds at altitude on any westbound flight. Taking these two facts into account, the actual range of the aircraft has now been reduced to 537 nm.

Another performance figure of importance is the service ceiling. For the Skyhawk this figure is 13,1000 feet. Since some of the higher peaks along the way reach above 14,000 feet, we simply can't plan to fly a straight line. The shortest route would seem to be Omaha, Denver, Las Vegas, San Diego. So let's take a closer look and see what that entails. Though the terrain between Omaha and Denver rises from 1,000 feet to 5,000 feet msl, this leg involves no problems. The distance is 420 nm which is easily within the range of the aircraft. But just beyond Denver we suddenly face that 14,000-foot wall of rock and from then on the flight is anything but routine.

Arriving in Denver in the early afternoon, there is plenty of daylight left, especially in the summer. To add a respectable number of miles to the day's travel is tempting. However, the prudent pilot will decide otherwise. During the afternoons and early evenings thunderstorms build in the Rockies with impressive rapidity, often obscuring mountaintops. And heat-generated turbulence is at its nerve-wracking worst. So the intelligent decision would be to stay in Denver, have a swim and a leisurely dinner, go to bed early and be ready to take off the next morning as soon after dawn as possible.

There are good reasons for starting early. The mountains are most beautiful during the early morning and late afternoon hours. The winds are likely to be relatively calm and last evening's thunderstorms will have dissipated, leaving just a few harmless puffs of cloud here and there.

The distance from Denver to Las Vegas is 542 nm which is more than should sensibly be attempted without refueling, especially considering that it involves long stretches of hostile real estate where airports are few and the distances between them long. The intelligent thing to do is to plan to refuel at Grand Junction and then to go on to Las Vegas.

It's 194 nm from Denver to Grand Junction as the crow is said to fly. Except we're not crows. The way we'll have to plan our route is considerably longer. We cannot sensibly expect our airplane to climb to an altitude which would take us safely across the 14,000-foot mountains west of Denver, not to mention that it would require oxygen which we don't have on board. The northerly and somewhat shorter detour involves Victor 8, leaving Denver in a northwesterly direction. This necessitates an immediate climb to 12,000 feet in order to get over even the lowest pass, which means that it might take a few S-turns or even 360s to get up there before getting too close to the mountains. This is important because, with the probability of westerly winds, we're likely to encounter downdrafts close to the eastern slopes. Visually, the easiest to follow landmarks are a railroad, road, and mountain stream, all heading straight west toward Corona Pass. But watch out! The railroad disappears into the seven-mile-long Moffat Tunnel. Just northwest of the mouth of that tunnel is Corona Pass, and once we're through that pass we again pick up a railroad and road (the same railroad, different road) which then lead us northwestward toward Kremmling and the Colorado River. The river flows from there to Grand Junction and can easily be followed. Once through that pass we'll also be able to receive the Kremmling VOR.

But much of the time the Kremmling area is covered by clouds, the bases of which frequently obscure the higher terrain. Therefore, unless the weather at Kremmling is reported to be strictly VFR, it might be better to plan for the southern and somewhat longer route.

76

This involves leaving Denver southwesterly on Victor 89 and then picking up Victor 95 at the Lake George Intersection. Following this route we have all kinds of time to get up to altitude, but eventually here, too, a 14,000-foot mountain ridge lies across our path. At that point Victor 95 ceases to be of value to us, which is just as well because we most probably have lost reception from any of the referencing VORs. There is a railroad and road combination that intersects V-95 at Buena Vista. This can be followed south. At Poncha Springs the road, accompanied for a while by a railroad and the Arkansas River, heads west toward and eventually through Monarch Pass. By following it we can stay at a reasonable altitude and, once through the pass, the Gunnison VOR will come in. We can then use it or the Tomichi Creek to Gunnison and the Blue Mesa Reservoir which can be seen for miles.

From Gunnison it's a no-sweat flight to Montrose which, on this route, should take the place of Grand Junction as a fuel stop. Making allowances for the various detours, the distance flown will be about 220 nm, and after fueling the airplane, defueling the pilot, and a cup of coffee it should still be amply early in the day to continue.

Now a decision must be made. Using Victor airways (V-244 to V-8 to V-8N) offers the advantage of continuous contact with appropriately placed VORs. Except for the Bryce Canyon area it is scenically pretty dull, so, if scenery turns us on, the thing to do is to leave Montrose on the 240-degree radial and fly to Canyonlands National Park (90.4 nm), then take up a heading of 220 degrees and fly over the entire length of Lake Powell which brings us at its western shore to Page (113 nm). From there we can follow the Colorado River south through Marble Canyon toward the Grand Canyon, and then we head west over the Grand Canyon to Lake Mead and Las Vegas (between 220 and 260 nm). The Victor Airway route totals 388 nm while the scenic route adds up to approximately 435 nm, give or take a few.

Now the sensible thing to do is to quit for the day. Any pilot new to the Rockies should by now have his fill of scenic splendor coupled with gradually but constantly increasing turbulence. Although the mountains ahead can't hold a candle to those behind, late-afternoon turbulence over the rest of the route will get a lot worse rather than better.

On the charts the flight from Las Vegas to San Diego doesn't look like much. The highest mountains along the way only go up to 7,500 feet and the while thing should really be a piece of cake.

It isn't. The distance via Goffs, Twentynine Palms, Thermal and Julian VORs is only 235 nm, but much of it is desert where the heat of the day produces wind and turbulence which can be upsetting, to pick a rather unfortunate phrase.

Again, if we can tear ourselves away from Las Vegas slot machines, we should leave at the earliest possible hour. If not, just tighten the seatbelt, relax and pretend it's a rollercoaster. A word of warning: Non-instrument rated pilots should be sure to check the San Diego weather. Low clouds and fog are a frequent occurrence and it would be depressing to have come all this way and then not be able to land at the destination.

The total route which has been described here includes some of the most rugged and, at the same time, beautiful terrain to be found in the country. The problems mentioned are typical of those which might be encountered when flying anywhere in the mountains. But simply planning the route carefully, knowing the altitudes at which to fly and the most convenient fuel stops is only one portion of what's necessary to be safe. What follows is a kind of mountain-flying checklist.

Aircraft and Instruments

The importance of range and service ceiling has been mentioned above. It might be added here that the service ceiling is not the absolute ceiling of an airplane. It is the altitude at which the aircraft, at gross, will still climb at 100 fpm. It will continue to climb at a constantly decreasing rate until a combination of full power, the right mixture, and a flight attitude approximating the best angle of climb will produce only level flight. (As an example, I once coaxed a Comanche 250 to an altitude of 22,600 feet, and it was still climbing at 10 fpm).

Rate of climb is another parameter that takes on greater importance in the mountains. Normally most of us climb at what we consider a comfortable rate and angle, little concerned about the time and distance it takes to get to the desired level. In the mountains, flying up toward a ridge or the saddle of a pass, we have to know with certainly that we can get up there in time, which means knowing how many feet per minute we can squeeze out of that airframe-and-engine combination under the prevailing temperature conditions. Pilots flying light twins should remember that most are incapable of maintaining the necessary altitude on one engine, and in the event of an engine malfunction, the most immediate problem will be to somehow head toward lower ground.

Takeoffs and landings on warm or hot days at high altitude airports are something that must be experienced to be believed. The effects of density altitude have been described in greater detail elsewhere in this book.

In addition to the usual navcom equipment, an ADF can be helpful. It is not at all unusual to be out of reception distance of VORs for long stretches at a time, but there is almost always a standard broadcast station somewhere to help give us some idea of our progress.

An EGT takes an increased value when virtually all flying is done at altitudes which require intelligent mixture control.

One instrument rarely found in light aircraft but worth its weight in gold, especially in the mountains, is an angle-of-attack indicator. It is simply a needle that indicates whether the attitude of the aircraft is right for climb, cruise, or descent relative to the airspeed, and it warns of approaching stalls. Mountain flying involves getting misleading cues from sloping terrain and horizons, and one is easily seduced into thinking that the attitude of the aircraft is different from what it really is. Flying the needle of an angle-of-attack indicator eliminates all that. Not being affected by density altitude it produces reliable indications at all times.

Another item to be considered is oxygen. Quite aside from the legalities involved, prolonged high-altitude flying without supplementary oxygen may result in severe headaches at best or, at worse, result in a loss of efficiency and increasing degrees of disorientation. In other words, it's just plain stupid. Though a planned mountain flight may not include altitudes above 12,500 feet, one never knows what conditions might be encountered. A narrow cloud bank obscuring a ridge (Fig. 12-1), uncomfortable turbulence,

Fig. 12-1. A narrow cloud bank obscuring a ridge may cause us to climb to oxygen altitudes.

or even just continuing updrafts may cause us to climb into oxygen altitudes, and, for this reason, at least a small portable oxygen system should be seriously considered.

Weather

Weather in the mountains is different. Visibilities are usually considerable, 70 miles or better being not at all unusual. Haze and fog conditions are rare, and the industrial pollution enveloping so much of the country is absent here. As a result we tend to underestimate distances. We'll look at a mountain or other landmark which seems quite near and then wonder why it takes forever to get there.

Thunderstorms build regularly during the afternoon hours in summer and fall (and sometimes in spring and winter). They rarely combine into squall lines but usually rise to impressive heights in solitary splendor. As elsewhere, they produce vicious winds, rain and hail, and must be avoided at all cost. Circumventing them is rarely a problem, assuming there is ample fuel on board for the additional miles resulting from the detour.

Small, fast-moving weather systems may develop between reporting stations and, therefore, remain unannounced. Try to circumvent rather than over- or underfly them. Overflying a cloud deck of any consequence can turn into a sucker trap (Fig. 12-2). From below the tops tend to look reasonable enough, but once up there, they often continue to rise and rise until they exceed the climb capability of the airplane. Then what? We can either turn back (unless, of course, the clouds behind us have climbed too) or file instruments. Or we can simply and illegally punch through, hoping to come out in the clear on the other end. But besides being illegal, it's more stupid here than in flat country, because we can't come down if we find that we have to. We must always be aware that the lower clouds may just be full of rocks (Fig. 12-3). Conversely,

Fig. 12-2. In the mountains, overflying a cloud deck can be a sucker trap.

Fig. 12-3. We must always be aware that the lower clouds may be full of rocks.

underflying such a cloud deck, unless it's distinctly clear of the ridges ahead, can also result in a dead end, literally as well as figuratively. In situations where clouds obscure mountaintops but a pass ahead seems clear, don't fly through the pass unless we know that there's a way out of the next valley or that there is an airport on which we can land.

Wind

Wind direction and velocity are affected by the configuration of the mountain ranges. We have talked about that in a separate chapter about wind.

Night

Forget it. Unless and until you're an experienced mountain pilot, don't fly after dark. Nothing is as black as a moonless night in that clear hazeless air over endless stretches of uninhabited rock. By the time the mountain ahead comes into view it may be too late to avoid it.

In summary, mountain flying can be a breathtaking and satisfying experience, but it is not to be taken lightly. Sloppy technique, slipshod preplanning, or inadequate maintenance of the aircraft, its systems and instruments can easily make a first mountain flight the last. But planning, caution, respect for the terrain and for its weather keep mountain pilots flying year in and year out, and there's no reason for proficient pilots from other parts of the country not to join their number.

In closing, any aircraft owner who expects to have to cross the Rockies with any degree of frequency would be well advised to consider installing a turbocharger and a permanent oxygen system. Then he can simply climb up to an altitude which keeps him safely above all obstructions.

Chapter 13

Winter

Depending on where we live, as much as half of the year may be winter; and if we want to keep on flying, some special precautions are necessary. Every phase of flight requires a degree of special attention: Airport conditions, preflight, in-flight, arrival, all take on added dimensions. Knowing what to expect and how to combat the conditions that are encountered will help make winter flying safe and enjoyable.

Airports

During the summer there are over 12,000 airports and landing strips in the United States available to the general aviation pilot. But during the winter quite a few of these are inoperative because snow and ice on the runways and taxiways make it impossible to use them with an adequate degree of safety. For this reason it is not only important to know the surface conditions at the departure airport before starting to taxi, it is also vital to be aware of the conditions expected at the destination airport at the estimated time of arrival. Furthermore, as conditions have a habit of changing rapidly and often unexpectedly, it is essential to carry at least double the fuel reserve considered ample during the summer months, as an acceptable alternate airport may be a long distance from a destination that has suddenly gone sour.

Tiedowns

Subfreezing temperatures cause oil to harden and batteries to weaken. They also tend to deflate landing gear struts if the aircraft flew in from a warmer climate. Strong winds may make it impossible to preheat the engine with external heater-blowers, and frequently the only way to get going is to have the FBO move the airplane into a heated hangar to get it thawed out. But hangar space is hard to come by during the winter and it may be necessary to make arrangements for it in advance to avoid lengthy delays.

Snow, Ice, and Frost

Snow, ice, or frost, even in small amounts, when adhering to the fuselage, wings and tail surfaces tend to deform the airfoil to a degree that may make takeoff and subsequent flight dangerous if not impossible. The aircraft must be clean to fly. Dry snow can be brushed off with a broom or rags, but frost and ice must be melted away, usually making it essential to tow the airplane into a heated hangar and leave it there until all traces of moisture have dried. Remember that moisture, once hit by the cold outside air, will instantly freeze to metal surfaces, thus defeating the purpose of having moved the airplane into the hangar in the first place.

Fast Changing Conditions

During the winter, weather systems and fronts tend to be fast moving and smaller in area than during the summer. As a result, weather conditions along the route of flight and at the destination may change drastically from hour to hour. A continuous listening watch should always be kept to be informed as to the current conditions ahead. Especially in mountain areas, clouds often form quite suddenly (Fig. 13-1), obscuring ridges and rolling down into the valleys and canyons, making continued VFR flight chancy if not impossible.

Icing conditions, usually a greater hazard to IFR pilots than to those operating VFR, are discussed in detail in another chapter.

Cabin Heat

Be sure that the cabin heater is in good working condition and that there are no leaks in the heater hoses. Sitting in an airplane with fingers and toes turning to ice can be a ghastly experience for pilot and passengers alike. But when using the heater always keep some ventilation going in addition to the cabin heat. Carbon

Fig. 13-1. Clouds often form quite suddenly, obscuring ridges and rolling down into the valleys.

monoxide has neither taste, smell, nor color, and it is difficult to be absolutely certain that the exhaust manifold doesn't have a leak somewhere that could cause exhaust gases to be sucked up by the heater and channeled into the cabin. Ideally the hot air should be directed at the feet, and some cold air at the face.

After Dark

Winter nights are longer than winter days. Therefore, winter flying without flying at night doesn't get us very far. In some respects VFR night flying is easier when the sky is clear and the ground is covered with snow. Roads are clearly visible as dark bands in the white landscape even when there is no moon. Also, unlighted airports (and when there is snow some runway lights may be buried under it) appear as dark strips against the white.

Preflight

The preflight weather check should, during the winter, include questions with reference to the expected conditions of the destination airport at the estimated time of arrival. If forecasts suggest the possibility of snow, freezing rain, or a drastic drop in temperature, even if conditions are expected to be VFR at the ETA, it might be advisable to contact the destination airport by phone to make sure that the runway will be kept clear of snow and ice. If there is any doubt, one should be prepared to use an alternate airport, preferably a major airline terminal that can be expected to have continuous clearing operations.

The next step is the actual preflight. The usual walk around should include a look inside the wheel-pants on fixed-gear aircraft, to make sure that no accumulations of mud or ice have frozen inside. Check fuel caps, vents and static ports and be sure that they are free of dirt or frozen moisture, or there may be the danger of collapsing a fuel tank.

Prior to starting, pull the prop through a few times to loosen the oil and thus reduce the strain on the battery, which puts out less power when it is cold. Under conditions of extreme cold an APU may be unavoidable. The oil will simply be too thick and the battery too weak to produce the degree of rotation necessary to get the engine to fire.

If the aircraft is equipped with a constant-speed prop, exercise the prop more often than during the warmer months. It takes a while for the oil in the mechanism to warm up and permit it to operate. Do not attempt to take off until at least the oil-pressure gauge is in the green. On really cold days the oil-temperature gauge may never get into the green until after the aircraft is airborne.

Taxi and Takeoff

Prior to releasing the brakes, be sure to know the condition of the taxiway and runway surfaces. Frozen moisture, patches of ice, and even small amounts of snow on the ground can make taxiing difficult and may affect takeoff considerations. Ice on the ground may make a runup impossible. The brakes won't hold the aircraft in place and it may be necessary to run up the engine while the aircraft is still tied down, making sure that no debris is being blown on other aircraft behind it. Or one might have to perform a cursory runup while taxiing and, if neither is possible, we may just have to take off without a runup.

If the runway is slippery and there is a crosswind, be prepared to slide sideways across the runway until adequate speed has been attained. Start the takeoff run on the upwind side of the runway to avoid sliding off the paved area.

En Route

Once airborne, the degree of necessary vigilance depends largely on the prevailing weather. If the visibility is good and the temperature way below freezing, there is nothing much to worry about. Just listen to the hourly sequence reports to be alerted early of any sudden change in conditions along the route. Aside from staying clear of clouds, know the types of precipitation forecast for your route. Dry powder snow, like rain, while reducing forward visibility, represents no particular problem. Freezing rain or wet snow, on the other hand, will adhere to the aircraft and windscreen, reducing forward visibility and sometimes, when heavy, covering the windscreen completely to a degree which is beyond the capability of the defroster to deal with.

Always continue to keep track of conditions at the destination airport. Remember, when braking action is being reported as being medium or poor by a car or truck, it is likely to be worse for an airplane touching down at much higher speeds. In that case, are the runways long enough to permit a safe rollout without using brakes?

Landing

The thing to worry about is the condition of the runway. If it's dry and clean, fine. If it's slippery, partially or wholly frozen over or covered with a thin film of snow, land short and slow and don't attempt to use the brakes until much of the speed has been dissipated. If, in addition, there is a crosswind, land on the upwind side of the runway and be prepared to slide sideways once the airplane starts to slow down. If several inches of snow cover the runway, especially if it is damp snow, keep the nose way up and come as close as possible to making a full-stall landing. The sudden braking action as the wheels plow through the snow could put more strain on the nosegear than it can take.

Watch for snow piled high by the sides of runways and taxiways. These piles of snow tend to get rock hard and, especially with low-wing aircraft, you might just end up with a dented wingtip.

Maintenance

In addition to normal maintenance, the following should be checked before every flight:

- Proper inflation of landing-gear struts.
- No leaks in heater hoses.
- Is the oil in the aircraft of the right weight for the temperature conditions in which you will be operating?

If, of course, you expect to be shuttling between New York and Florida or Idaho and Southern California, you may just have to suffer the delays and expense involved with preheating the engine, because using too light an oil in warm weather is more harmful than too heavy an oil in cold.

To the knowledgeable pilot willing to take reasonable precautions, winter flying in light aircraft can be as safe and enjoyable as at any other time of the year. And when the weather is clear, flying over the snow covered expanse of the plains states or the intimidating white peaks of the Rockies can be breathtakingly beautiful.

Chapter 14

Ice Is for Cocktails

The pilot of the Comanche had been the victim of his own optimism. Impatient with scooting around under a low broken overcast and detouring around intermittant showers he had climbed to VFR conditions on top and now found himself at 12,500 feet above St. Louis. The trouble was that the broken clouds had long ago coalesced into a solid deck. The other trouble was that he was due in St. Louis within the next few hours. Lambert Field was reporting 800-foot overcast with three mile visibility in light rain. Well, instrument ticket or no instrument ticket, he'd either have to file and then descend through this stuff, or he'd have to go elsewhere and forget about getting to St. Louis on time.

Oh well, what the heck. He felt confident that he would be able to make the approach under the reported conditions, so he called St. Louis flight service and filed IFR to Lambert Field. A short time later he was told to expect vectors to an ILS approach to Runway 12R.

"Radar contact. Descend to and maintain 6,000. Report passing through 10 and eight. Report reaching six. Maintain present heading." The pilot acknowledged, reduced throttle, and set up a comfortable rate of descent. By the time all the blue sky and the sun above had disappeared (Fig. 14-1) he had successfully forced himself to stop paying attention to the outside and to, instead, concentrate on his instruments. He felt certain that he would be able to follow ATC instructions and that, with an 800-foot ceiling

Fig. 14-1. Soon the sky above and the sun were disappearing.

and three-mile visibility he would be able to land safely even though he had never actually flown an ILS approach.

The first indication that something wasn't quite the way it should be came when he found that he had difficulty receiving ATC. The transmissions kept breaking up and several times he had to ask for three or more repetitions to make sure that he had understood correctly what he was being asked to do. He made a mental note to have his radio checked the first chance he got.

The second indication was more subtle. Somehow the airplane didn't feel quite right. It seemed to be in an awful hurry to lose altitude and it didn't seem right to have to add power in order to maintain the desired rate of descent. And holding altitude required nearly full power with an uncomfortably nose-high attitude.

Though based in Southern California where icing in clouds is not a routine occurrence, the pilot had, by now, figured out that he must be picking up ice. Never having had this happen to him before, he was unaware of the rapidity with which ice tends to form when the temperature and humidity situation is just right, and he, therefore did not tell ATC what was happening.

As it turned out, he was relatively lucky. ATC cleared him to descend from 6,000 feet without much delay and once passing through 4,000 the outside air temperature climbed to comfortably above freezing, and by the time he broke out of the clouds with

Runway 12R right over his nose, the airplane was behaving normally again, and the radio reception was as it should be. Once stopped on the ramp, there wasn't an indication anywhere of the amount of ice which the airplane must have been carrying only a few minutes earlier.

An Iced-Up Plane

It is too bad that pilots rarely get an opportunity to actually see what an iced-up airplane looks like in flight. No spoken or printed warning could ever expect to scare us the way the real thing would. In this particular incident the initial reception difficulties were caused by ice buildup around the VHF antenna. This used to be a lot more serious with the older types of antennas. The newer and more streamlined variety seems to be less prone to ice accumulation.

The airplane itself tends to attract ice wherever there are protrusions such as rivet heads, metal seams and such. And ice likes to adhere to ice, so as soon as a tiny bit has formed near one of those metal irregularities it is likely to grow and spread rapidly. The result with reference to the aircraft's performance can be quite drastic. Ice is heavy and the kind we are talking about here usually results in a rough surface which plays havoc with the airfoil. Thus, as more ice adheres, the angle of attack must be increased in order to maintain level flight. This exposes more and more of the underside of the airplane to the slipstream which usually is less clean than the upper surfaces and thus prone to accelerate the ice accumulation. Being in areas which the pilot cannot see, his only means of guessing what is happening is the way the aircraft handles.

Visible Moisture

For ice to be able to form in the first place there must be what is referred to as visible moisture, in other words clouds, freezing rain or haze with an extremely high humidity content. The outside air temperature most conducive to ice formation is from about a degree or two above freezing to eight or 10 degrees below freezing. Once it gets colder than that the moisture is already frozen and is not likely to adhere to the aircraft. As we all know, temperatures vary with altitude. Usually it gets colder the higher we go, but under certain conditions it might be colder near the ground and warmer higher up. Those of us who are expert at reading weather charts can figure out which situation is likely to occur where. But the fact is that most of us are not expert at reading those charts, not to mention that when we need them they usually don't happen to be handy.

The reason for wanting to know what is happening temperature-wise above and below is the need to make a decision quickly, climb or descend, when we have gotten ourselves into an area where ice is forming. What we're looking for is either a level of warmer air which will permit the ice to melt, or much colder air which is most often found in clear VFR conditions above. Once in the colder air the ice will gradually dissipate through evaporation. In making such an up or down decision we must also take under consideration the ability of the aircraft to climb speedily to the desired altitude with whatever load of ice has already accumulated, plus whatever additional amount it will pick up during the climb. If the aircraft is one which, under these conditions, is still able to achieve an honest 500-fpm rate of climb, fine. If, on the other hand, the best that can be coaxed from the airplane is a dawdling 150 or 200 fpm, it might be better to start looking for some place lower down, because the airplane is likely to arrive at a point at which it will barely maintain level flight before it gets to the adequately cold altitude above.

VFR pilots are, of course, not supposed to find themselves in icing conditions. But what is supposed to be and what actually does happen are often two different things. It is, therefore, important to at least know where ice might be found and what to do when it is found.

Types of Icing

So far we have talked about icing in relation to the airframe and to aircraft performance. There are other types of ice which, while not exactly in the same classification, do deserve to be talked about.

Carburetor Ice. Somehow this never seems to have been particularly well understood and today with an increasing number of engines being fuel injected, it is likely to become even more obscure. First of all, carburetor icing is possible in the middle of summer on a hot, humid day. It has little if anything to do with the outside air temperature. What happens is that the fuel-air mixture, while undergoing whatever it is that the carburetor does to it, cools quite rapidly to such a degree that if there is moisture present, it is likely to freeze. If there is little moisture, the tiny particles, frozen or otherwise, simply move through the engine with all the rest of the fuel mixture, never to be heard from again. But if there is a lot of moisture, such as when the temperature and dew point are only a few degrees apart, then this moisture can freeze into clumps adhering to the intake orifices, resulting in a slow but sure choking of the engine. So, whenever a gradual unexplained reduction in the power output of a carbureted engine is noticed, pull the carburetor heat all the way out. Unless the pilot has waited too long, the carburetor heat will melt the ice and after some spitting and coughing near-normal power output will be regained. It won't be full power, because the preheated air will alter the fuel-air relationship and reduce the amount of power available. To minimize this power reduction the engine should be re-leaned if prolonged operation with carburetor heat is anticipated. The correct routine is to continue to fly with carburetor heat full on as long as the conditions of excessive moisture remain present. Never use partial carburetor heat; like being pregnant, it's all or nothing.

Frozen Rain. Another ice-related problem is one which is rarely serious, but can be quite unpleasant. Here is what can happen: You take off in the rain, maintain a steady rate of climb and eventually arrive in clear conditions above the clouds. You now try to trim the airplane for level flight, except it won't trim. The trim tab, soaked from the rain, froze as you climbed to a below-freezing altitude and now it's stuck, and it is not going to come loose again until you get back down to where it's warmer. Solution? Either you have to fly the entire way to your destination pushing like mad on the yoke in order to keep the nose down, or you've got to throttle back to some sort of slow-flight speed which will keep the airplane level in that nose-high attitude. Or you have to go back down until the thing comes loose and then climb back up, exercising the trim tab and, possibly, the other controls repeatedly in order to avoid having them freeze on you again. There are extreme cases on record where the primary controls, elevators, rudder and ailerons, froze as the result of a similar situation. If that should happen the only sensible solution would be to throttle back and let the airplane settle back down until the controls start to work again. A variation on this theme would be frozen flaps. This could be scary, especially in an airplane with hydraulic or electric flap actuators, because they might bend or tear something, possibly ending up with one flap down and one up, a situation which as been known to kill people.

Iced Pitot Tube. Still another icing condition to watch for is what can occur when there is moisture in the pitot tube and we climb to below freezing temperatures. The moisture will freeze and all airspeed indication will be lost. If the aircraft is equipped with pitot heat, that will usually take care of it, and pitot heat should always be used as a preventive measure under such conditions. If no pitot heat is available, there is nothing that can be done. We simply have to fly without an airspeed indication which is actually not as difficult as it might seem. Similarly, when flying an aircraft equipped with one of those automatic gear-lowering devices which are operated by air pressure into a pitot-tube-like gadget, the air intake can freeze over and suddenly the gear will come down unannounced. Unless we're cruising at an excessively high speed, this is not likely to do the gear any harm, but the gear doors, being more fragile, could be bent or actually be torn off.

It should not be necessary to state that taking off in an airplane with any amount of ice, snow or even frost on the wings, tail surfaces or fuselage is extremely dangerous. One wonders how often those famous last words, "Oh, that isn't enough to mean anything," have been spoken a short time before a fatal and final crash. Admittedly, it is a nuisance to drag an airplane into a heated hangar to let it thaw out and dry off or to spend a half hour or more in freezing wind with a broom trying to clean away the last remnants of last night's snow storm, but that's just the name of the game, and if we want to play it safe, that's what we better do.

Chapter 15

The Concrete Behind

It used to be standard procedure for ground control to issue taxi clearances to the end of the active runway unless the pilot requested an intersection takeoff. Now, with more and more two-mile-long jet runways, light aircraft are frequently given taxi clearance to the most convenient intersection even without a request from the pilot. For all practical purposes nothing has changed. While it used to be up to the pilot to request the intersection, it is now up to him to request the full length of the runway. The difference is psychological. Requesting an intersection takeoff makes the pilot feel that he is telling the controller that he is a good enough pilot to get his bird safely airborne in less than the full length of the runway. On the other hand, requesting the full length of the runway seems to imply a lack of proficiency, especially when the runway in question is one of those long ones (Fig. 15-1). As a result pilots are prone to accept intersection takeoffs even though they may realize that they are taking a calculated risk.

How Much Is Enough?

The question that few lightplane pilots are able to answer with certainty is: How much runway is enough? We all know, or at least should know, the minimum field length which is required to clear a 50-foot obstacle. This figure is listed in the owner's manual. But do we always remember what this figure really means? It means

Fig. 15-1. Requesting the full length of the runway tends to make the pilot feel that he is admitting to a lack of proficiency.

that, at gross, flown by an expert test pilot, with a brand new engine, the aircraft is just able to get off in time to clear that obstacle at a sea-level airport when the runway is smooth, the temperature is 59 degrees F and there is no wind. But what about when the temperature is 85 degrees, the runway is rough, the field elevation is 4,000 feet and there is a gusty crosswind? And how much additional runway would we need if the engine suddenly starts to act up just as we are about to lift off, and we have to shut everything down, roll out and come to a stop without running off the end of the runway?

Balanced Field Length Figures

The performance figures for jet aircraft, instead of listing minimum field lengths, include so-called balanced-field-length figures which represent the distance necessary to accelerate to lift-off speed and then come to a full stop. This is a much more meaningful figure but it is not readily available for light aircraft. Some manufactuers have recently started to include an accelerate-stop distance in the published performance parameters, but this is still fairly rare. As a general rule it is safe to assume that the accelerate-stop distance is approximately twice the 50-foot-obstacle-clearance distance, more if the runway is wet.

This may seem like a lot of talk about nothing. After all, most of us would have a tough time remembering when we last aborted a takeoff. But that doesn't mean that it might not become necessary tomorrow. The reasons for at least considering aborting a takeoff are legion:

- A bug crawled into the pitot tube and there is no airspeed indication.
- Just at the moment of liftoff a door pops open.
- While accelerating the pilot dropped his glasses and they slipped down behind the rudder pedals.
- The door was inadvertently closed on a seatbelt and it's now hanging outside and banging against the fuselage.
- An engine malfunction or actual failure.

And let's not fool ourselves into thinking that just because we're flying a twin, we always have the other engine to complete the take-off, go around and then land again. Half the light twins on the market don't have sufficient single-engine climb capability to get the aircraft over the trees past the end of the runway.

It's something to think about the next time we're cleared for an intersection takeoff. How much runway is left and how much do we need to accelerate and then stop, just in case one of those bugaboos shows up just as the wheels are about to lift off the concrete. So the controller thinks you're an old fuddy-duddy because you're asking for twice as much runway as you'll actually be using. So what? Let him think what he wants. It's not him that's up for grabs. It's you. Always remember that old cliche: There is nothing more useless than the sky above, the fuel on the ground, and the concrete behind.

Chapter 16

Greyout

The pilot was on his way from the East Coast to Detroit. He had flown from New York more or less straight west and had reached the southern shore of Lake Erie somewhere west of the city of Erie and east of Cleveland. Despite a fairly high overcast, the weather had been fine so far with, for this part of the country, pretty good visibilities. Ever since the lake had come into view he had been arguing with himself about whether to fly across the lake on the straight route toward Detroit, or whether to stay over dry land and fly the detour via Cleveland and Toledo. The difference in distance and, therefore, time as well as fuel would be considerable. From his present position, straight across the water, would involve about 100 nm, while flying around the lake would nearly double that. The weather for Detroit was reported as 3,000-foot overcast, visibility five in smoke and haze. Cleveland was reporting 6,000-foot overcast, visibility four. In other words, if he flew across the water the overcast would probably be lowering somewhat along the way, but it shouldn't be any serious problem except that having to fly about 2,000 feet above the water (Fig. 16-1), he'd be out of sight of land and, of course, out of gliding distance for some time.

Oh, what the heck. The airplane doesn't care whether it's flying over land or water, so why should he? He tuned his OBI to the 286-degree radial from the Jefferson VOR which should take him past Windsor, Ontario, on the south and straight toward Detroit. And now, keeping the needle centered, he headed out over the water.

Fig. 16-1. Having to fly at about 2,000 feet, he would be out of sight of land and gliding distance for a long time.

Everything Was Fine

At first, everything was fine. He was flying at 4,500 feet and watched the lake shore gradually disappear behind him. He did wish that he would be able to see the horizon more clearly, but everything in the distance seemed to dissolve into some sort of featureless grey. There was no wind to speak of and the water below was also calm and without distinctive features except for a freighter to his left, steaming eastward. He had been over the water for about 10 minutes when it became obvious that the ceiling was beginning to press down on him and he dropped down to a lower altitude, leveling eventually at 2,000 feet. With the surface of the lake being at around 575 feet, this put him at about 1,400 feet agl. At this lower altitude whatever wavelets there were below could be seen more clearly, but straight ahead and in all directions all he could see was grey water merging into grey haze and grey sky (Fig. 16-2). He was aware that even though it was certainly technically VFR, he was, in fact, on instruments and he wished that he had a wing leveler or autopilot which didn't need a horizon in order to keep the aircraft straight and level.

Time Passed Slowly

In such a situation time has a way of passing much more slowly than normal. Time and again he would look at his watch, only to

Fig. 16-2. In all directions all he could see was gray water.

realize that only a minute or two had passed. No matter how hard
he tried, there was simply nothing his eye could fasten on except
the water below (Fig. 16-3) and even it seemed so calm and flat
as to be meaningless. By now the overcast was not too far above
him and he wondered what he would do if it forced him down even
lower. The trouble was that the overcast was just a solid sheet of
grey and he couldn't even be sure where it was, except that he knew
he was not in it. If it did drop down, would he be able to see it before
he found himself suddenly in the clouds? After all, with everything

Fig. 16-3. No matter how hard he tried, there was nothing his eyes could fasten on.

around him being just one shade of dirty grey, how would he know if what he was flying into was simply more of the same old haze or actually some lower clouds?

He remembered having read about pilots flying over Greenland and being caught in a whiteout where actually everything around them was white, and there was at least one story about someone having landed in the snow without realizing it. Well, at least he still had the water to look at and the OBI . . . or did he? He suddenly realized that the CDI needle had remained so firmly centered not due to his faultless navigation but rather that it had ceased receiving the signals from the VOR. He simply had failed to notice the OFF flag. He tuned the nav receiver to the Windsor VOR, but to no avail. He remembered noticing the minimum transmission altitude for his route to be listed as 2,400 feet on the Jeppesen chart, and he figured, probably correctly, that he must be somewhere in the middle of the lake where, at his low level of flight he'd be beyond reception distance for either station. But, after all, there wasn't much chance of getting lost. Even if he should inadvertently drift off course, he'd have to reach one of the shores sooner or later, and with the water below barely moving he had to assume that there wasn't enough wind to be of consequence. Well, he'd just have to sit there and be patient and hope that the ceiling wouldn't play any tricks on him.

Again minutes seemed to be stretching into hours, but eventually there was a shudder in the CDI needle, the OFF-flag popped back and forth and then it disappeared for good and the needle settled slightly to the right of center. Apparently he had done pretty well in terms of holding his heading, because he did want to pass Windsor to the south, meaning that the needle would be to the right as long as the OBI was set to his heading rather than the actual radial from Windsor on which he was at any given moment.

Then There Was A Ship

And then there was a ship, and then another, and after a few more minutes he thought he saw some sort of a dark line way ahead in the distance. The shore? He continued to peer into the haze and the dark line got darker and more distinct. True enough, it was the shore and finally there was something for him to look at again. He breathed a sigh of relief, surprised that the appearance of an indication of dry land should make him feel so much better.

Actually it all had been perfectly simple. Nothing to it. Whatever there had been that had been bothersome had been his own apprehensions. Granted, if the engine had decided to act up, or if there had been some other malfunction, he might have been in some fairly serious trouble. People were probably right when they said that one shouldn't fly across any of the Great Lakes except on clear days when one can climb to 10,000 or 12,000 feet, thus reducing the time beyond gliding distance to shore to an acceptable minimum. But, anyway, there was the shore, Canada to the right, the good old U.S. to the left and straight ahead, and along with it the smoke and air pollution which is one of the distinguishing marks of Detroit. All he would have to do now is find Detroit City Airport and he'd be home free.

Chapter 17

Instruments

In many of the chapters in this book much emphasis is placed on the need to control the aircraft by reference to instruments alone. The reason is simple. More often than not, when a VFR pilot gets himself into a marginal-weather situation, he will sooner or later find that he has lost visual contact with outside reference points, forcing him to rely on his instruments in order to return to VFR conditions.

Not too many years ago the training involved in obtaining a private license included no instrument training at all. In those days flying by reference to instruments was actually considered a common student error. The student was taught to control the aircraft attitude by reference to the horizon and instructors often covered the instruments in order to force the student to use whatever outside reference was available. He was taught to respond to what he saw, heard, and felt in the seat of his pants.

In more recent years, basic flight training has started to include a certain amount of instrument work under the hood. But whether or not those few exercises are sufficient to last a pilot through years of VFR flying is questionable. The trouble is that the difference between contact flying and instrument flying is largely psychological. While in contact flying the feel of the airplane is a vital source of information to which the pilot reacts more or less automatically, in instrument flying he must, above all, learn to ignore that feeling because it is based on faulty reactions of the senses to given flight situations (Fig. 17-1).

The human body is simply not designed for flying. It is designed to operate on the ground and the sensory system is programmed to react to fixed reference points. It is difficult to adjust the habits developed since early childhood with reference to movement on the ground to the new habits required for instrument flight without recognizing the reasons why the old habits must be consciously ignored. The primary sensory system which tends to give us trouble is the Eustachian tube in the inner ear which controls our sense of balance. Once deprived of constant correction supplied by visual cues, this sense of balance turns into the equivalent of a used-car salesman in south Los Angeles: It tells us what it thinks we want to be told, not what is actually fact. Thus, once we are in a bank for even a very brief period of time, the inner ear adjusts to that condition and tries to persuade us that the bank is, in fact, straight and level flight. If we then pay attention to that sensation and we want to keep on turning, we'll automatically increase the angle of bank only to, after a moment, be told again that we are flying level. Unless corrected by reacting to what the instruments show, the pilot will soon find himself in a screaming spiral.

On the other hand, when we level off after having been in a bank for a few moments, the inner ear will try to convince us at first that we are, in fact, banking in the opposite direction. Again, only the instruments are able to provide us with a true picture of the attitude of the aircraft.

Fig. 17-1. Once all visual contact with outside reference points is lost, the pilot must learn to ignore seat-of-the-pants sensations and rely solely on his instruments.

The instruments used for this purpose are referred to as air-data instruments and they are the altimeter, the airspeed indicator, the vertical-speed indicator, the turn-and-bank indicator, and the artificial horizon plus, of course, the magnetic compass and directional gyro. Using these instruments as sources of reliable and accurate information on which to base our control of the aircraft requires that we have at least a smattering of knowledge of how they work.

Altimeter

The altimeter reacts to changes in atmospheric pressure (Fig. 17-2). Its reading is only correct when it has been set to the prevailing atmospheric pressure at the location at which the aircraft is being operated. If we have set the altimeter to the atmospheric pressure at our departure airport which, at the time we left, was located in a high-pressure area and read, say, 30.26, and we have since flown into a low-pressure area where a barometer on the ground would read something like 28.87, the altimeter in the aircraft will be reading high, meaning that it will tell us that we are considerably higher than we actually are. This may not be

Fig. 17-2. Altimeter. Courtesy Mitchell Aircraft Instruments, Inc.

105

particularly important when we know that we're several thousand feet above the highest obstacle anywhere in the vicinity, but it can become critical at low altitudes. It is, therefore, of great importance, especially if we are operating in marginal weather conditions which, more often than not, are associated with low atmospheric pressure, that we check with the FSSs in our area of flight to ascertain the correct local altimeter setting.

Airspeed Indicator

The airspeed indicator reacts to air pressure in the pitot tube. Its reading is referred to as indicated airspeed (IAS) and differs from true airspeed (TAS) because the thinner air at altitude produces less pitot pressure than does the heavier air near sea level. Since the primary purpose of flying for most people is to get from one place to the other in a hurry, we like to convert the IAS reading to TAS, because it always is the higher figure, thus making us feel that we are moving with greater speed. Many modern airspeed indicators have a rotating dial which can be set to read the TAS at any given altitude (Fig. 17-3). Failing that, the TAS can easily be figured by using the E6B flight computer, or one of the new electronic pocket computers.

In practice, though few of us may be willing to admit this, TAS is really of little importance. It is useless in determining the actual ground speed unless we happen to know the head- or tailwind component at our altitude and position. Neither can it be used in determining how far above stall speed we happen to be operating. The stall speed of an aircraft is given in IAS, not TAS, and it changes neither with altitude nor with temperature and the resulting density altitude (though it is affected by aircraft weight and possible deformation of airfoils due to icing or such). This is important to remember because marginal weather conditions frequently cause us to want to climb rapidly or make steep turns, either of which can cause us to inadvertently get dangerously close to a stall.

When deprived of visual reference to outside cues, the airspeed indicator is an important instrument in keeping us posted as to the attitude of the aircraft. When the nose is raised above the horizontal plane, airspeed will drop. When the nose is lowered, airspeed will rise. This is especially important during slow flight or flight at extremely high altitudes where we tend to be operating at an excessively high angle of attack while actually remaining in level flight. In the event of a failure or malfunction of the suction system

Fig. 17-3. An airspeed indicator with provision to show true airspeed based on altitude and outside air temperature. Courtesy Mitchell Aircraft Instruments, Inc.

which operates the artificial horizon and the gyro compass, the airspeed indicator is the only remaining instrument to judge flight attitude, unless the aircraft is equipped with an angle-of-attack indicator (an extremely useful but little-appreciated instrument).

Vertical Speed Indicator

The vertical-speed indicator (VSI) tells us the vertical speed, in feet per minute (fpm) at which the aircraft is either climbing or descending. It is useful during prolonged climbs and descents, but its indications tend to lag somewhat and, therefore, are not reliable immediately after a climb or descent has been initiated. It is helpful when we want to trim the airplane for a change in flight altitude at a specific fpm rate, but is best thought of as a secondary instrument.

Turn-and-Bank Indicator

The turn-and-bank indicator or needle-and-ball (Fig. 17-4) is useful primarily in helping us make coordinated turns and preventing us from slipping or skidding. It can also be helpful in setting up a continuous standard-rate or similar turn, though most pilots prefer to use the artificial horizon for this purpose.

Artificial Horizon

The artificial horizon is, without a doubt, the primary instrument used when controlling the aircraft by reference to instruments alone. It displays the position of the wings with reference to the horizontal plane (horizon), it is calibrated to show the degree of any bank, and it displays the position of the nose of the aircraft with reference to a level plane. In its display the aircraft symbol is fixed while the horizon line moves and banks. Initially students, being psychologically conditioned to think of the horizon as a fixed line and the aircraft as something that moves with reference to that fixed horizon, often find it difficult to interpret

Fig. 17-4. Turn-and-Bank Indicator. Courtesy Mitchell Aircraft Instruments, Inc.

Fig. 17-5. Magnetic Compass. Courtesy Mitchell Aircraft Instruments, Inc.

the readings of the artificial horizon in a hurry. But a bit of training and experience quickly teaches us to read this instrument instantaneously. Though it is possible (and a part of actual instrument training) to control the attitude of the airplane by needle-and-ball and airspeed alone, it is considerably more difficult and the untrained VFR pilot should be warned against taking a chance of getting into instrument conditions in an aircraft which is not equipped with a working artificial horizon.

Magnetic Compass

The magnetic compass (Fig. 17-5) and the directional gyro are two instruments which must be used in combination with one another. The directional gyro is the primary instrument in maintaining a given heading or turning to a specific heading. But all directional gyros will gradually precess in one direction or the other and they must, from time to time, be compared with the magnetic compass to make sure that their indication is correct. This comparison must be accomplished while in straight and level flight as even slight turns, climbs, or descents will cause the magnetic compass to produce an inaccurate reading.

Instrument Use

the VFR pilot without adequate instrument training should make a conscious effort to relax when entering instrument conditions. Excessive tension will cause a pilot to overcontrol which can easily result in what is commonly referred to as unusual attitudes. Probably the best advice is to take the hands off the yoke altogether and keep the aircraft straight with just a bit of rudder pressure whenever the instruments show that it is starting into a shallow bank, and to keep it in level flight by use of the trim wheel. In this manner the aircraft, operating at normal cruise speed, can be kept in level flight and in a wings-level attitude indefinitely (or at least as long as there is fuel in the tanks). Somehow we are less likely to use excessive pressure on the rudder pedals than we are when we try to correct the flight attitude with our hands on the yoke. The other advantage is that even if too much pressure is exerted on the rudder pedals, the reaction of the airplane tends to be quite minor and it is pretty nearly impossible to put the airplane into an unusual attitude by use of the rudder alone.

Unless there is a special reason, such as proximity to the ground or some obstacle, the pilot will probably be better off by spending a few minutes with his hands in his lap, flying the aircraft just with the rudder and trim while getting mentally adjusted to the fact that he will have to rely solely upon his instruments in order to eventually get back to VFR conditions.

There can be no hard and fast rules for the next step. It depends entirely on the conditions which exist and how aware the pilot is of the weather conditions in all directions within range. If you have good reason to believe that by continuing straight ahead at your current altitude you will sooner or later break out of the clouds, continue on using just feet, and simply relax and wait. If a turn to the right or left or a complete 180 is indicated, then you will eventually have to use your hands because it would be fairly difficult to try and accomplish a major change in flight direction by use of the rudder alone. Pay no attention to anything outside the airplane. Keep the eyes on the artificial horizon and don't grab the yoke with your fists but just touch it lightly with the tips of the fingers. Turn just a little at a time and watch the position of the fixed aircraft symbol with reference to the moving horizon. Keep the dot firmly placed on the horizon line and place the wing symbols on a one- or two-dot bank but no more. As soon as the desired bank has been established, the ailerons may have to be neutralized in order to prevent the bank from increasing. Depending on the direction of the bank, it may even require opposite aileron to maintain the desired angle. The aircraft will probably have a tendency to drop its nose slightly and it will require a light amount of backpressure to keep the dot on the horizon line. Don't overcontrol. Very light control inputs are all that are needed.

Once the bank has been established, glance over at the directional gyro (which should have been checked against the magnetic gyro before the turn was started) to see how much longer you'll have to stay in the turn until the desired heading is reached. Remember that even though your instruments tell you that you're in a bank, the inner ear, your sense of balance, will try to convince you that you're straight and level. Ignore it! As soon as the new heading is reached, use a slight amount of aileron pressure to return the wings to a level attitude, and then continue on the new heading using just your feet and the trim wheel as before.

Climbs and descents are best accomplished by strictly keeping the hands off the yoke and using the trim wheel and throttle. At low altitudes, where normal cruise speed is attained at less than full throttle, simply advancing the throttle may accomplish an adequate rate of climb without the need to adjust the trim. At higher altitudes where full throttle is normally used for cruise, only the trim will need to be adjusted. In aircraft with constant-speed props, don't fool with the rpm. Leave it alone. Except in extreme circumstances the additional climb capability which can be obtained by increasing the rpm is too minimal to worry about.

Descents, at any altitude, can usually best be set up by simply reducing the throttle setting a fraction, and without bothering with the trim. Remember, the less you fool with the controls, the less the chance of getting into trouble because of overcontrol or faulty control application.

If at all possible, try to avoid climbing or descending turns. If both a climb or descent and a turn are indicated, do one at a time. In the case of a climb and turn, climb first to put all possible altitude between you and the ground, and then, once level at the higher altitude, start the turn. In the case of a descent plus turn, make the turn first while there is still lots of room between the airplane and the ground. Then, when the new heading has been established, start the descent.

All this can, of course, be practiced under a hood in VFR conditions, but not when you're alone in the airplane. Someone should be in the right seat, preferably, though not necessarily, another pilot. He will have to look out for traffic while you are busy keeping your head in the cockpit. If it's another pilot, he'll know what to do. If it's a non-pilot friend, point out to him where to look for other traffic. There's no point in his getting upset about another airplane way up above you or way below. Instruct him carefully in what to look for where, and then sneak a look yourself from time to time, just to make sure that he's doing his job. And do such practicing at a safe altitude, preferably 5,000 feet agl or higher, and away from heavily traveled airways or busy student-training areas. But do practice. It may some day save your life.

Chapter 18

Wing Levelers and Autopilots

Ever since Mooney started to produce its aircraft with a wing leveler as standard equipment about a decade ago, the number of weather-related accidents involving VFR pilots flying Mooney aircraft has dropped noticeably. The fact is that the cause of most of such accidents, beyond the obvious flying into instrument conditions, is the inability of the VFR pilot to maintain control over the aircraft by instruments. Basically all aircraft though relatively stable will eventually start to drop one wing, and this initially shallow bank then gradually increases to a steeper and steeper angle, unless the pilot uses the controls to correct for the situation. Not so if the aircraft is equipped with a wing leveler or autopilot. Even the simplest form of wing leveler will keep the wings level for an indefinite period of time and will re-level them if the pilot has forced the aircraft into a bank. In addition, all have means of initiating a shallow turn which maintains a constant shallow bank angle. With these capabilities in the aircraft, the VFR pilot is relieved of much of the complication associated with controlling the aircraft by reference to the instruments. He can, instead, concentrate on the best way back to VFR conditions, can study his charts and communicate with Flight Service Stations on the ground without worrying that the airplane will suddenly do him dirty.

Prices

Simple wing levelers are available at reasonable prices for single-engine aircraft.

An optional feature that is highly recommended automatically navigates on a given bearing or radial to or from a VOR. Considering the cost of airplanes and just about everything associated with them, this is about the cheapest insurance a VFR pilot can buy.

Drawbacks

The one negative aspect of flying a wing-leveler or autopilot-equipped aircraft is that it may increase the temptation to fly into instrument conditions because much of the danger factor has been eliminated. It must always be remembered that any time a VFR pilot flies into IFR conditions without communicating with ATC there is the danger some legitimate IFR traffic is operating in the area and that he is endangering not only his own life and that of his passengers, but also the lives of those in the other aircraft.

It is unavoidable in a book of this type that we must talk frequently about actions which are not only illegal but also dangerous to the occupants of the aircraft involved and to other innocent aircraft which may be operating quite legally in the same airspace. It can't be emphasized too frequently or strongly that such illegal operations in instrument conditions should be considered only as a last resort. Any rogue pilot who habitually ignores the FARs and blithely operates on instruments without bothering to obtain an ATC clearance is, in fact, a menace to aviation in general and to himself in particular.

Chapter 19

Knowing Where You Are (and Getting Where You Want to Be)

The willingness of the VFR pilot to fly on top of an overcast, at night, or under any other conditions which preclude visual contact with ground-based landmarks should be tempered by his ability to use his navigation equipment with precision. Just because every licensed pilot has once upon a time been taught how to use his VOR and maybe even the ADF equipment, this does not necessarily imply that he'll be able to locate himself precisely over a specific point on the ground if that point is hidden from view. Virtually all of us can recall incidents when we knew perfectly well that we were close to a certain airport where the conditions were 1,500- or 2,000-feet scattered or broken with three miles visibility in haze, and we had one heck of a time finding it.

Let's go through a quick refresher course on the use of the various types of navigation equipment available in the average better-equipped aircraft today. What we'll be discussing is the basic VHF navigation receiver and the different faces of the associated OBIs, DMEs, ADFs with fixed and rotatable compass roses, simple area navigation equipment, HSIs and lastly Loran C.

VHF Nav Receivers and OBIs

The purpose of the VHF nav receiver (Fig. 19-1) is to receive signals from VORs and to translate these signals into a cockpit display which, when correctly interpreted by the pilot, shows the direction of the aircraft's present position from the VOR. This

Fig. 19-1. Nav receiver, bottom, and OBI, top. Courtesy Narco Avionics, Inc.

cockpit display, generally referred to as the omni-bearing indicator (OBI), is not affected by the direction of flight of the aircraft from the station (Fig. 19-2). In order to use this equipment to determine the distance from the station as well as the direction it is easiest to tune in two VORs and determine the exact position by triangulation. While this can be done with one receiver and one OBI by switching back and forth between the frequencies for two stations, it is a lot simpler if two receivers and two OBIs are available (Fig. 19-3). Since an airplane cannot stand still in the air, the time lapse associated with switching from one frequency to another and then readjusting the OBI will materially degrade the precision with which we can determine our position.

An alternate method of determining how far we are from the station, the only available method when only one station is within reception distance, is to fly at right angles to a given radial and watch how long it takes to cross radials which are 10 degrees apart. The formula for this procedure looks like this:

- Time in seconds between bearings (radials) divided by the number of degrees of bearing (radial) change equals minutes to the VOR.
- True airspeed (or ground speed, if known) times the time in minutes between bearings (radials) divided by the degrees of bearing (radial) change equals miles to the station.

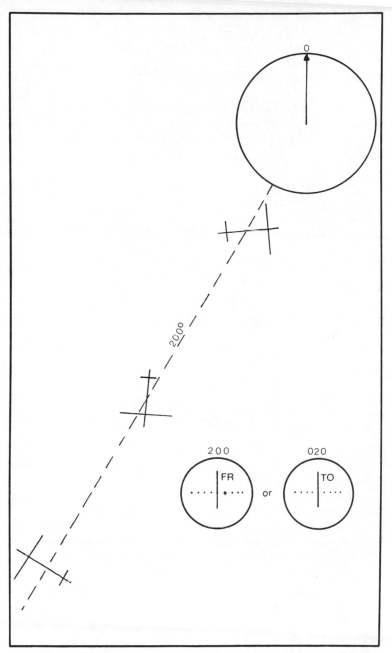

Fig. 19-2. The OBI shows the radial or bearing on which the aircraft is located. It is not affected by the distance from the station or the direction of flight.

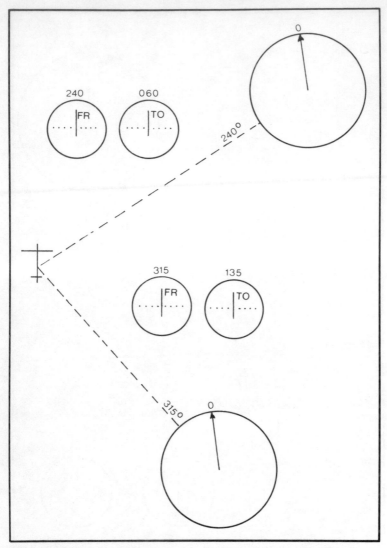

Fig. 19-3. The exact position of an aircraft can be determined by the process of triangulation, using two VORs.

If it takes the aircraft three minutes to fly a 10-degree bearing change, for example, the aircraft is 18 minutes from the station, assuming that the ground speed remains unchanged: 180 seconds/10 degrees equals 18 minutes.

Using the second method, if the aircraft is flying at 150 knots (or mph) then the distance to the station is 45 nm (or sm) from the

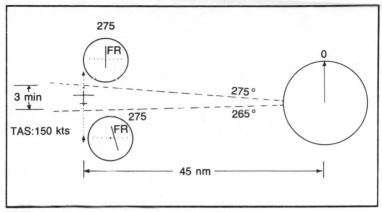

Fig. 19-4. If it takes an aircraft flying at 150 knots three minutes to cross 10 degrees of radials, the distance to the station is 150 × 3 ÷ 10 = 45 nautical miles.

station (Fig. 19-4):

$$\frac{150 \text{ knots} \times 3 \text{ min}}{10° \text{ equals } 45 \text{ nm}}$$

This method of determining position relative to a station is cumbersome and, unless ground speed is known with any degree of precision, it is inexact—but it's better than nothing. Since most of us are likely to be unable to remember the formulae involved, it might be a good idea to put it on a card and keep it with the charts or in the glove compartment.

So far we have used the faces of conventional OBIs in the illustrations. In recent years several companies have manufactured OBIs with different displays. Some, instead of a swinging needle, use a vertical bar as the course-deviation indicator (CDI). This bar moves from side to side and is somewhat easier to interpret with precision than the swinging needle (Fig. 19-5). Another company, Bendix, has done away with both the needle and the bar, and has substituted horizontally lined-up short vertical electronic bars. The number of such bars being lit indicates the degree of deviation from the selected VOR radial (Fig. 19-6). In addition several manufacturers, following the trend established by Collins in its Micro Line, have added a radial and bearing readout capability to the nav receiver itself. With this display in the cockpit we know at all times the exact radial from or bearing to the station on which the airplane is located (Fig. 19-7). Having this information

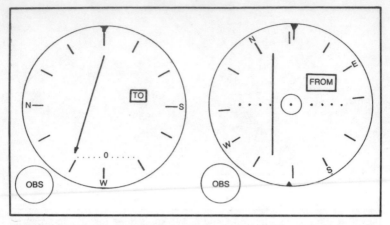

Fig. 19-5. Two faces of an OBI, one with a swinging needle, one with a moving vertical bar.

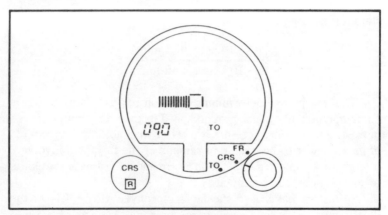

Fig. 19-6. The face of the OBI developed by Bendix Avionics for its 2000-series of navigation instrumentation.

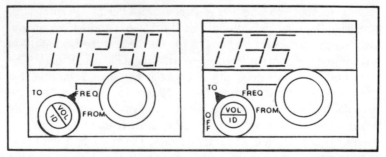

Fig. 19-7. Additional information displayed by Collins' Micro Line navigation receiver.

constantly available greatly simplifies the task of navigating with precision.

In order to avoid confusion about the different designations and abbreviations used in conjunction with OBIs, Fig. 19-8 shows what is meant by them.

While few VFR pilots are ever confronted with the need to utilize localizer frequencies, it should be understood that the conventional nav receiver and all OBIs are designed to receive and display not only information relative to VORs but also to localizers. A localizer is a directional beam used by IFR pilots in making instrument approaches to airports equipped with a full ILS or a localizer. Localizers are designed to provide horizontal guidance to the runway and their frequencies range from 108.1 to 111.9 MHz, all of them utilizing the odd-tenths frequencies only (108.1, 108.3, 108.5 etc.). When the receiver is tuned to a localizer frequency the OBS is inoperative and the CDI will react when the aircraft passes anywhere through the area covered by the localizer. When the aircraft is on the localizer centerline, the CDI will be centered, when

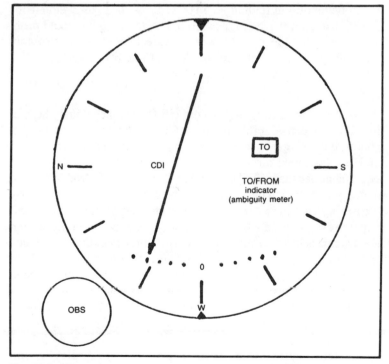

Fig. 19-8. The various components of an omni-bearing indicator.

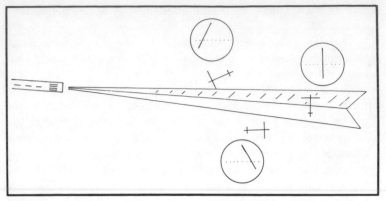

Fig. 19-9. Indications of the OBI when crossing or flying in the reception area of a localizer.

it is to either side of the centerline, the CDI will be off center in the direction to which the aircraft must be flown in order to intercept the centerline (Fig. 19-9). Under ordinary circumstances the ability to receive localizers is of little use in navigation, but occasionally it can be helpful in finding an airport. Since localizer frequencies are not listed on the average aviation charts, the pilot would probably have to call the airport tower or nearest FSS in order to obtain that information, or refer to the Airport/Facility Directory.

DMEs

Distance measuring equipment (DME) is marvelously helpful in adding a considerable degree of precision to the navigation chores, but, being quite expensive, it is not often found in the lighter and less-equipped aircraft. DME (Fig. 19-10) is so-called pulse equipment because, similar to radar, it transmits pulsed signals to a ground station which are then returned to the aircraft. By measuring the time it takes for each such pulse to make that round trip, the system is capable of figuring out how far from the station the aircraft is located. Then, by measuring the rapidity with which this distance changes when the aircraft is flying either directly to or directly from such a station, it can figure out the ground speed of the aircraft. Most DMEs are designed to display either distance to or from the station in nautical miles, the ground speed in knots, or the time to or from the station in minutes, assuming no change in the current ground speed. Some will display only one of those parameters at a time while others are designed to display several or all of these data simultaneously.

The obvious advantage of having DME is that only one station need be within reception distance in order to determine the exact present position. And this information is constantly available to the pilot without the need of bothering with cumbersome triangulations. In addition, by providing ground speed information on a continuous basis, the pilot can compare that information with the TAS and know precisely what wind component is present at his altitude and location. Having this information in front of us simplifies the task of selecting an altitude at which we can take advantage of the greatest tailwind, or conversely, find the flight level with the least headwind.

But DMEs only function in conjunction with so-called co-located VOR/DME facilities or with VORTACs. They cannot provide the information referenced to a simple VOR. This reduces the number of stations available for navigation somewhat, but there is an

Fig. 19-10. A combined VOR, ILS, DME and marker beacon display unit with DME (digital numbers) in the range mode. Courtesy Narco Avionics, Inc.

adequate number of VOR/DMEs and VORTACs in most areas of the country to assure reception as long as the aircraft is at some reasonable altitude. Figure 19-11 shows the symbols used on aviation charts to differentiate between simple VORs, VOR/DMEs, and VORTACs.

ADFs

Automatic direction finders (ADF) are probably the oldest type of electronic navigation equipment and though today relegated to a position of secondary importance, they still do come in handy. (Fig. 19-13). While VOR receivers are designed to tell us the location of the aircraft relative to the station regardless of the direction of flight, the ADF display shows us the direction to the station relative to the nose of the aircraft. There are two basic types of ADF displays. One has a fixed compass card on which zero always coincides with the nose of the fixed aircraft symbol. The other has a compass rose which can be rotated by the pilot to place the heading at which he is currently flying under the nose of the fixed aircraft symbol. The latter is a lot easier to use because it relieves the pilot of the chore of mentally figuring out what the needle deflection means in terms of actual direction from the aircraft to the station (Fig. 19-13).

Most of the time the average VFR pilot will use his ADF to fly to a station and to know when he crosses over that station. He will not, and probably should not, attempt to use it in order to navigate away from a station as this requires considerable practice.

ADFs operate on frequencies between 190 and 1799.95 kHz which include non-directional beacons (NDB), compass locators (LOM), and standard AM broadcast stations. When tuned to any of these facilities, the needle will point toward the transmitting station and the pilot can navigate toward that station by simply keeping the needle right under the nose of the fixed airplane symbol. But the frequencies in this range are affected by rain, sleet, snow, thunderstorms, mountains, coastlines, disturbances in the upper atmosphere, day, night, dusk and dawn. The farther away from the station and the higher the frequency, the more serious the errors are likely to be. During daylight hours, when the weather is halfway decent, ADF reception is usually quite reliable. During dusk and dawn it is notoriously unreliable. During the night it is less reliable than during the day, but more than during dusk and dawn. In addition, even though the frequency allocations for NDBs and standard broadcast stations are made to theoretically avoid interference

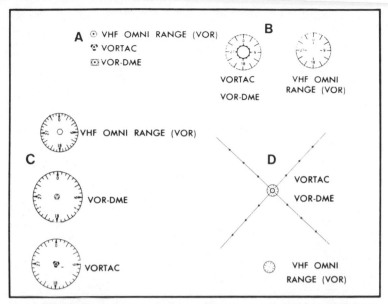

Fig. 19-11. Symbols used on aviation charts to describe VORs, VOR/DMEs and VORTACs. (A = sectional and WAC charts, B = Jeppesen low-altitude en route charts, C = NOAA low altitude en route charts, D = Jeppesen area navigation en route charts.)

Fig. 19-12. A digital ADF tuner with display head and rotatable azimuth card. Courtesy Narco Avionics, Inc.

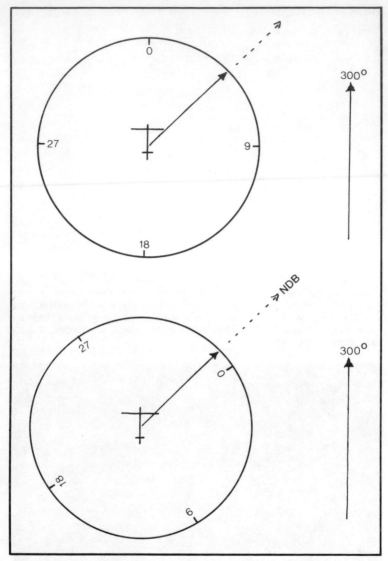

Fig. 19-13. The difference in the ADF display between units with stationary and rotatable compass roses.

with one another, there are times when such interference can be produced by stations located at an incredible distance. (Once, while flying at night from Phoenix to Los Angeles, I found myself listening to perfectly clear reception from a standard broadcast station which turned out to be in Philadelphia). It is important, therefore, to always make certain that the station being used is, in fact, the station we think it is. This is easy with NDBs and other low-frequency nav aids which broadcast an oral identification signal. It is more complicated in the case of standard broadcast stations which often identify themselves only once every half hour or so, and then not necessarily exactly on the hour or half hour.

It is important to know if the needle of the particular ADF installed in the airplane has a regular park position, a position to which it moves when no adequate signal is being received. It could simply stop moving, thus giving no visual indication that it is no longer pointing to any particular station. Unless the station-identification code is being monitored, flying according to the ADF needle could result in drastic navigational errors. The sound is usually a tip-off. If it is filled with static and cuts out intermittently, it is safe to assume that the signal is unreliable regardless of what the needle does. Conversely, if the sound comes in strong and steady without obvious interferences (even though the sound of most ADFs is pretty bad), it can be assumed that the needle indication is reliable.

With the ADF needle always pointing to the station and if that station is our destination or check point, the logical assumption would be to align the needle with the nose of the airplane, and then to keep it there until the needle flips 180 degrees which would indicate station passage. With no crosswind this works fine. In crosswind situations it would result in flying a curved track which, if the crosswind is of considerable velocity, might add quite a few miles to the distance being covered. In that situation the thing to do is to fly a heading and to apply the necessary crosswind correction (the slower the aircraft, the greater the correction), using the ADF needle as a backup rather than as the primary navigation aid.

As long as we fly *to* the station things are fairly simple. But they get a bit muddled when we fly *from* the station. The needle will point straight to the tail of the airplane no matter what direction we fly from the station. In this case we must fly a heading, using the needle to keep track of whether a crosswind is causing us to drift off course. When this happens the needle will begin to point several degrees to one side or the other of the tail. We then need to make a course correction in the direction of the needle. We should continue that turn until the needle indicates twice the number of degrees as before. We continue on that heading until the needle shows the same deviation as was shown before we started the correction. Then we turn back on course, keeping the needle a few

127

degrees off the tail in the direction from which the wind is blowing (Fig. 19-14).

In the U.S., where there is no shortage of VORs, the ADF is usually considered a secondary instrument except in the case of instrument pilots who use it in making ADF approaches. But once we go beyond the borders of this country we find that VORs are few and far between and here the ADF becomes our only means of navigating.

Area Navigation Equipment

Unless or until they have had an opportunity to fly with it, most VFR pilots will consider area-navigation equipment (RNAV) an expensive frill which does little to enhance the ease and safety of VFR flight. Nothing could be farther from the truth. The capabilities of RNAV, when fully and correctly understood and utilized will greatly enhance the ability of the VFR pilot to cope with the many weather and navigation problems which he is likely to encounter.

RNAV, in principle, provides the pilot with the ability to electronically relocate VOR/DME or VORTAC stations to any position within 100 to 150 nm of the actual station. To accomplish this the system utilizes the capabilities of an all-purpose digital computer in conjunction with a conventional nav receiver, OBI and DME. The nav receiver is tuned to a given VORTAC and the DME is tuned to that same VORTAC. The computer is then told at what distance from the VORTAC and on what radial from the VORTAC the pilot wants to establish the new phantom VOR which is referred to as a waypoint (W/P). The interface between the pilot and the computer is the so-called control-display unit (CDU) (Fig. 19-15). It is equipped with pushbuttons or concentric knobs to enter digital data into the computer, plus a display window which shows the pilot the coordinates of the waypoint in question. When the RNAV system is activated and a waypoint has been selected, the CDI needle in the OBI operates in relation to that waypoint just as if it were an actual VOR, the only difference being that the needle deflection is linear, meaning that a deflection of a given number of dots always indicates a given number of nautical miles off course, rather than a given number of degrees (Fig. 19-16). And the DME will display the nautical miles to the waypoint, ground speed, or time to the waypoint at the current ground speed.

One of the most obvious and basic examples of using RNAV is in the context of finding an airport which has no nav aid associated

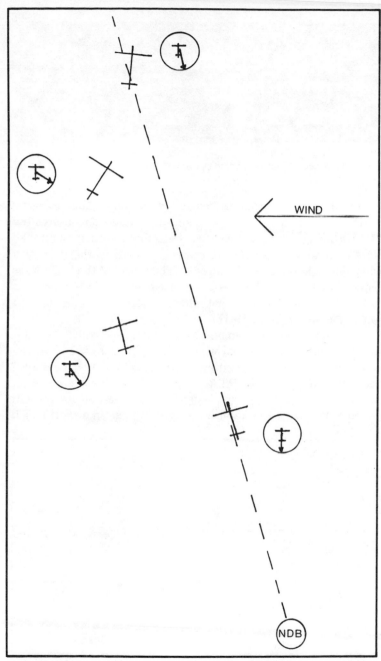

Fig. 19-14. The ADF display when flying from the station and correcting for wind drift.

Fig. 19-15. A typical area navigation system (RNAV). Courtesy Narco Avionics, Inc.

with it. The pilot checks his charts for the radial from a nearby VORTAC on which that airport is located, and he measures the distance along that radial. He then feeds that information into the RNAV computer and, presto, there is a VOR in the form of a waypoint right on top of the airport. Then, even if the visibility is reduced by haze, smoke, smog or what have you, he simply flies to that waypoint, knowing that when he gets there he is right on top of the airport (Fig. 19-17).

Now let's take that famous VFR situation in which we want to take a look into marginal weather conditions to see if we can go on or if farther on it's getting so bad that we may have to turn back. We now pick a VORTAC which is located some distance, say 30 to 40 nm, along our desired route ahead of our present position. We use it to establish a waypoint over an airport which

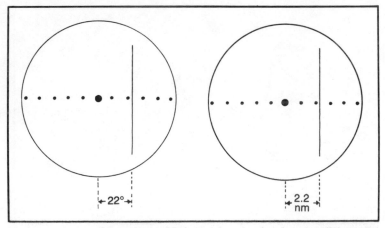

Fig. 19-16. In the VOR mode the CDI is angular, showing degrees off the radial (left). In the RNAV mode it is linear, showing nautical miles off the radial (right).

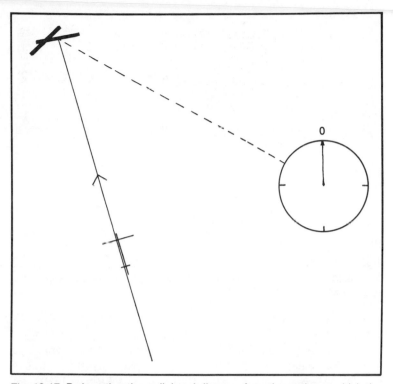

Fig. 19-17. By inserting the radial and distance from the station at which the airport is located into the RNAV computer, we can place a waypoint right on top of the airport.

is located in the clear somewhere near the point where the weather begins to get bad. Now we can fly into the marginal conditions and, if we want to, continue on our course for as far as we can receive the VORTAC to which our waypoint is referenced. Depending on our level of flight, in this instance this would mean that we can go on for some 80 nm without losing the waypoint. If we then find that we simply can't go on, we can navigate back to the airport on which we have placed the waypoint, find it without difficulty and then land (Fig. 19-18).

Or, let's assume that the winds aloft are such that it is important for us to know constantly how long it will take us to get to our destination, which may be 200 nm distant. This is a little trickier. We select a number of VORTACs more or less along our desired course and establish a number of waypoints all at the same location, namely our destination. By doing this we will get a constant DME readout with reference to our destination telling us, for instance,

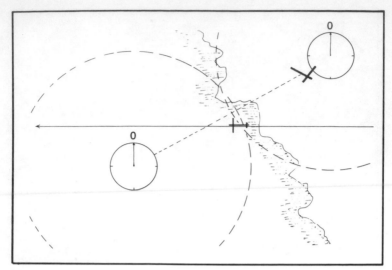

Fig. 19-18. RNAV can be useful in providing a safe escape route when the weather ahead begins to turn sour.

whether or not we have sufficient fuel to get there with the prevailing headwind component (Fig. 19-19).

The potential uses of RNAV are so varied that it would be impossible to describe all of them in these pages. Suffice it to say that RNAV greatly enhances the ability of the VFR (and IFR) pilot to navigate with precision, regardless whether he is flying along established airways or on a direct route along a course which is

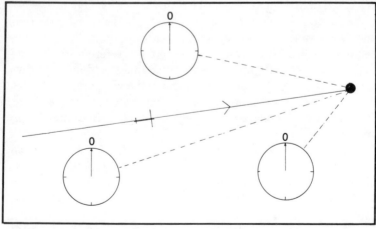

Fig. 19-19. By placing a series of waypoints over the destination, the DME will always read distance and time to the destination.

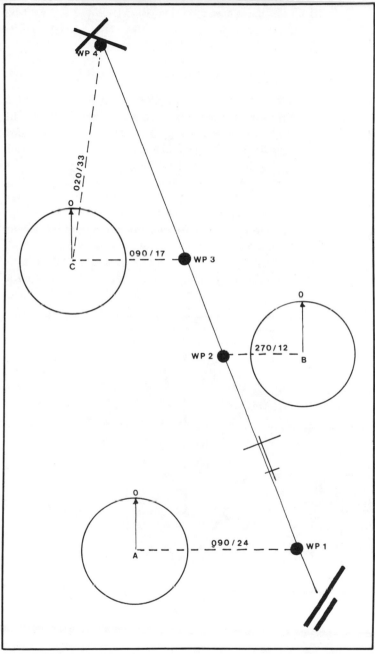

Fig. 19-20. If we want to fly direct off-airways, we can move the existing VORTACs to more convenient locations.

not defined by ground-based nav aids (Fig. 19-20). It takes a bit of practice, experience and imagination to utilize RNAV to its fullest.

HSIs

Horizontal situation indicators are, in fact, a more sophisticated version of an OBI (usually also including a glide-slope display). The average HSI consists of a directional gyro (usually slaved), a heading bug, a CDI, a TO/FROM indicator and a stationary aircraft symbol, plus a glide-slope indicator (Fig. 19-21).

The DG is just like any such vacuum instrument with a rotating compass rose which, when related to the stationary aircraft symbol in the center, indicates in which direction the nose of the aircraft is pointing. The term slaved means that the annoying precession problem is automatically corrected by a magnetic azimuth transmitter, also called a flux detector or flux gate. This is an instrument affixed to the aircraft in an area free of magnetic disturbances. It senses the alignment of the aircraft with respect

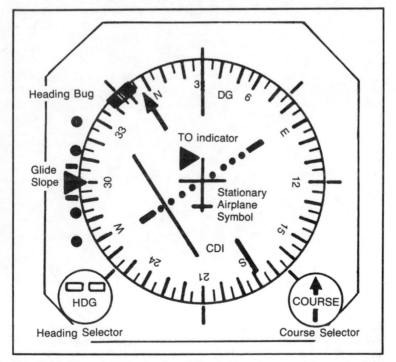

Fig. 19-21. The components of an HSI.

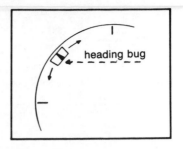

Fig. 19-22. The heading bug can be set by the pilot and coupled to the autopilot.

to the earth's magnetic field and sends that information to the DG, causing it to constantly align itself with the magnetic north.

The heading bug is hand set by the pilot to remind him of his intended path of flight. It can be coupled to an autopilot with heading hold which will then automatically turn the aircraft to keep the heading bug aligned with its nose (Fig. 19-22).

The CDI performs the same function usually performed by the CDIs in OBIs. The needle shows the aircraft's position relative to a selected VOR radial or bearing. Only this presentation, once understood, is easier and faster to interpret. The CDI consists of an arrow with a movable center section. By turning the course-selector knob it can be turned to a desired bearing TO or radial FROM the station. If, when this is done, the center section moves to the left it means that the aircraft is to the right of the bearing or radial and must turn left in order to intercept it. If the center section moves to the right the opposite is true. By relating the fixed aircraft symbol to the movable center section of the arrow, the appropriate course corrections can never be in doubt (Fig. 19-23).

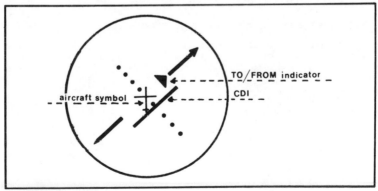

Fig. 19-23. The CDI is an arrow with a movable center section. In order to get back onto the radial, the pilot flies in the direction in which that center section is displaced (toward the needle).

135

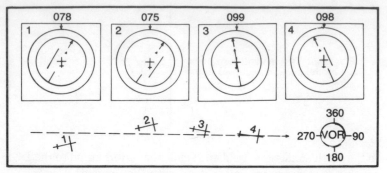

Fig. 19-24. The HSI indication while a pilot tries to stay on the 270-degree radial (90-degree bearing) from a station, flying in an easterly direction.

The TO/FROM indicator, also called ambiguity meter, shows whether the selected radial indicates the direction FROM or TO the VOR. It is usually a triangle resembling a stylized arrowhead rather than a flag reading TO or FROM.

The aircraft symbol in the center of the HSI is stationary and always points to the 12-o'clock position. When trying to interpret the instrument indications with respect to the aircraft, one might think of oneself as being above the aircraft, looking down. Figures 19-24 and 19-25 show the various HSI indications during a typical horizontal and a vertical maneuver.

Loran C

Loran C is probably the most revolutionary navigation aid since the compass. When a pilot enters a longitude/latitude coordinate,

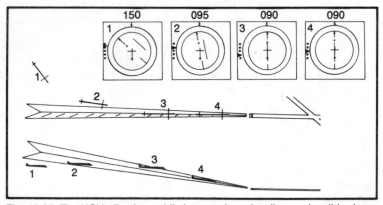

Fig. 19-25. The HSI indications while intercepting a localizer and a glide slope.

for instance a destination airport, the computer displays the heading, distance, and more information. Or ask the computer where you are and it will indicate the airplane's longitude/latitude; no VOR radials, no RNAV computations, and no DME. Reception is so reliable in most parts of the United States that Loran C is approved for limited IFR use. And, thanks to electronics miniaturization, Loran C receivers can be installed in the smallest airplane, even ultralights, or carried onboard the plane as a portable unit.

Since this book deals with flying in poor weather, rather than "everything you always wanted to know about Loran C," this brief section will be an overview. I would suggest TAB Books' *Flying with Loran C* by Bill Givens for additional information.

Givens puts it in perspective:

"Loran [an] acronym for *LO* ng *R* ange *N* avigation, [is] a method of measuring the position of an aircraft or boat by measuring the time difference between two sets of low-frequency radio signals propogated by a Loran chain station."

If an airplane is equipped with a functioning Loran C receiver and the pilot knows how to operate it, that pilot should be able to fly to practically *any* point he desires. If the weather turns sour and visibility drops to IFR conditions while en route, a pilot could lay in a course to good weather, fly by referring to the Loran C unit and proceed to his destination in VFR conditions.

It's been a whirlwind romance between Loran C and general aviation. Givens' book says VFR pilots were using Loran C in the late 1970s and the FAA approved the first helicopter installation in 1979 for IFR operations. The first airplane certification for IFR operations came in 1981 in a state-owned twin-engine Beech. And, stealing a cliche', the rest is history.

By early 1986 the Aircraft Electronics Association estimated between 27,000 and 28,000 Loran C units were flying in general aviation airplanes, according to *Business and Commercial Aviation*. The avionics industry believes demand will continue, bolstered by approvals for non-precision instrument approaches using Loran C and increased efforts to get full coverage in the Continental United States, the magazinc said. To get full coverage, a mid-continent "gap" where IFR accuracy is not assured would be eliminated by adding transmission stations. VFR operators may legally use Loran C in the "gap."

Regardless of its intended use, VFR or IFR, Loran C is great safety insurance and could pay for itself in fuel savings with direct flights. Fierce competition in the avionics marketplace, fueled by supply and demand, has driven the price down to reasonable levels. Getting acquainted with Loran C, and possibly acquiring a unit, could be a lifesaver in poor weather some day.

Further reading:

FLYING WITH LORAN C, Bill Givens, © 1985, TAB BOOKS INC.

Business and Commercial Aviation, © April 1986, "Avionics 1986."

Chapter 20

Flying to Air Shows

There isn't anything different about flying to the location of an air show except that, once we get close to our destination, there is often an amount of traffic which makes Chicago's O'Hare look like child's play (Fig. 20-1). Year after year the number of aircraft movements recorded at major air shows, as well as air races, glider and balloon meets, and similar aviation events exceed those which are normally handled by some of the busiest hubs in the country. The unprepared pilot attending one of these events for the first time may become unnecessarily intimidated as he listens to the never-ending radio chatter while still 30 or 40 miles from his destination.

Advance Preparations

As with virtually everything in aviation, a bit of advance preparation will tend to minimize the problems to be faced. Nearly all aviation events which can be expected to attract large numbers of aircraft are associated with some non-standard procedures, such as special frequencies, requests to use a specific approach procedure without communicating with the tower, or some such. These procedures are published in advance in the form of Notams and may, if devised early enough, also appear in the more popular aviation publications. While most of us, in the course of our normal flying activity, usually pay little attention to Notams, this is one instance when studying them in advance will pay dividends. Most

Fig. 20-1. Let's go to an air show!

FBOs are regular subscribers to the Notams and many will put the more important ones on their bulletin boards.

The following is a description of the procedure which was being used the last time I flew to the EAA fly-in at Oshkosh. Different procedures may be in effect during different years, but it illustrates why advance knowledge is of value.

When I was still some 50 miles out, too far to get any decent reception of the ATIS, I listened to Oshkosh Tower. There was relatively little chatter on that frequency and every few minutes a voice came on, saying over and over: "Aircraft arriving for the EAA fly-in, do not contact the tower. Monitor 124.5 (it may have been another frequency, I don't remember) and follow instructions." I then turned to that frequency where an apparently recorded voice gave these instructions: "Aircraft arriving for the EAA fly-in, do not contact the tower. Enter the downwind leg over the stone quarry northwest of the field and follow the aircraft in front of you. Monitor 118.7 (or some other frequency) for clearance to land. Do not transmit. You'll be cleared to land when turning final." This may not have been the exact wording, but basically it is what the instructions consisted of.

Not being familiar with the Oshkosh area, the first problem was now finding the stone quarry (Fig. 20-2). I headed for some imaginary point northwest of the airport and, sure enough, I soon began to see airplanes of all types and sizes following one another downwind and, by searching the area from which they came I eventually spotted that quarry. I headed for it and turned downwind behind a bright-red experimental and ahead of an Aztec (I was flying a Turbo-Centurion). Once established in a situation like that it is important to be proficient at slow flight because, more often than not, the airplane ahead may be doing just 60 or 70 knots, and overtaking him is strictly a no-no. By now, listening to the other frequency which had been broadcast earlier, I could hear one aircraft after another being given landing clearance, usually not by number but by type and color: "Green single-engine Cessna turning final, cleared to land . . . Yellow Cherokee behind the Green Cessna, continue approach . . . Red experimental, follow the yellow Cherokee . . ." and so on. It was a good system and worked just fine, the best part being that it eliminated the frustrating problem of getting a word in edgewise when the tower is excessively busy and you're inexorably coming closer and closer.

The last time I flew to the now defunct Reading Air Show the routine was different. Listening to ATIS we were told to contact Reading Tower on several different frequencies, depending on the direction from which we were coming. The trouble was that there were scattered clouds with bases around 1,000 feet or so, and the

Fig. 20-2. The first problem would be to find that stone quarry.

visibility below the clouds, though said to be three miles, was lousy to put it charitably. In addition to giving us the frequencies, the voice also stated that the airport would be closed to all arrivals and departures from such and such to such and such a time because of an airshow and, looking at the clock in the airplane (a Bellanca Super Viking) I saw that I had 30 minutes to get on the ground, or I'd be stuck with either hanging around in the air for several hours, or going elsewhere to land and wait.

For some silly reason, every time I flew to Reading VFR I had difficulties finding that airport, and I understand from others that I'm not alone. I tuned in the appropriate tower frequency and listened to the instructions being given other aircraft which usually amounted to ". . . downwind for runway 36 (or whatever), report abeam the tower". Somehow I didn't feel like calling them and announcing myself, not knowing how long it would take to find the airport. So I stayed above the scattered clouds where the visibility wasn't too bad and flew toward the airport, hoping that I'd see it once I got there.

Well, if I remember correctly, it was just about three minutes before the airport was supposed to be closed when I finally spotted the runway right beneath me between the clouds. "Reading Tower, Bellanca One Two Three Four Alpha, above the airport, with the numbers, landing."

"Bellanca Three Four Alpha, downwind for Runway three six. Report abeam."

I pulled back the throttle, dropped gear and flaps and banked sharply to see as much as possible beneath me, dropped down, managed to figure out which was the right runway and somehow got myself established on downwind. I reported, was told to follow an Aerostar turning base and finally got the clearance to land when I was about 100 feet from the threshold, maybe 10 or 15 feet above the ground.

Another example involves the time I flew to the air races at Cape May, New Jersey. I don't remember the exact routine being used at the time. What I do remember was being established on final when suddenly there appeared a twin below me and overtaking me. Well, that is enough to give you momentary heart failure. Quite obviously, he never saw me at all, and I only saw him when he was far enough ahead of me to pose no further problem, assuming I could slow down my airplane, a Cherokee Six, to a speed which would permit me to land far enough behind him to not run into him before he had a chance to clear the runway. It all worked out fine,

Fig. 20-3. A by-gone era. Tie downs at the defunct Reading Air Show in Pennsylvania, but this is still a good example of airplane parking at a major air show.

but in retrospect it is somewhat scary to think what would have happened if I had decided to steepen my descent while he was apparently directly below me.

In all of these situations, extreme vigilance is of major importance. There is no way for us to depend on tower controllers or whoever is handling the traffic on the ground to warn us of potential traffic conflicts. It's strictly a see-and-be-seen operation. In addition, it must be remembered that such aviation events usually attract a great number of pilots who may only have a few hundred hours in their logbooks and who may, therefore, be somewhat less efficient in heavy-traffic situations than the ones we usually encounter around busy hub airports. It turns out to be much like driving in traffic. We've got to not only fly our own airplane, we've got to be aware of everyone around us and be ready to take unexpected evasive action if someone suddenly makes an unexpected maneuver.

The Get-Away

The other half of the coin involves getting away from those places. When the fly-in, air show, air race, or whatever it is is over, usually on the afternoon of the last scheduled day, everybody wants to get out of there in one great fat hurry. Thousands of pilots start their engines and everybody wants to be among the first to taxi out to takeoff. Again the routine used by the controllers varies from location to location. Most of the time the tower and ground control are entirely left out of this process. Instead there are controllers on the ground waving flags who first funnel the airplanes from the

143

Fig. 20-4. Today's big show. Tiedowns at EAA's annual Oshkosh, Wisconsin fly-in. Courtesy Experimental Aircraft Association.

tiedown area (Fig. 20-4) to the taxiways and then to the takeoff end of the active runway. Here other controllers, also with flags, will wave one aircraft after the other off. In some places such as Reading, for instance, where the runway is wide enough two aircraft may be waved into position side by side and then may be given takeoff clearance in quick succession, with the aircraft on the right expected to make a right turnout, and the one on the left expected to turn left.

Occasionally, especially at locations where only one aircraft can take off at a time, the congestion is such that it may take as much as an hour or more to taxi out. In such instances, particularly on warm summer days, it is advisable to shut down the engine and to push the aircraft, in order to avoid having the engines overheat, not to mention the considerable fuel savings.

Considering the huge amount of traffic being handled in an unconventional manner, it is surprising that mishaps are a rarity. Most of the credit for this must necessarily go to the large numbers of pilots involved who seem to behave consistently with all possible consideration for the other guy. That then is the secret word: If everyone is considerate of his fellow pilots and uses more than the usual amount of vigilance, flying to air shows can be a safe and enjoyable experience.

Chapter 21

Tales of Five Rivers

The experiences related in the next few pages involve five different parts of the country, five different rivers, five VFR pilots, and five single-engine aircraft. They are based on actual flights or attempted flights between Boise, Idaho, and Portland, Oregon; between Seattle, Washington, and Portland; Cincinnati, Ohio, and New York City; Cedar Rapids, Iowa, and Memphis, Tennessee, and Charlotte, North Carolina, and Washington, D.C. In all cases marginal weather and a river played an important part.

The Snake River

The first river in our anthology is the Snake River, the border between Idaho and Oregon, west and northwest of Boise. The pilot had spent the night in Boise and had planned to fly to Portland the next morning. His aircraft was a Cessna Skylane equipped with dual nav receivers, a transponder and a simple Century I autopilot with VOR-coupling capability.

When he got up in the morning the weather in Boise was amply VFR despite a high broken overcast, occasional rain showers and strong westerly winds, gusting occasionally to 30 knots. His route would take him from Boise along Victor-500 via Kimberly to south of Portland where he would then turn north toward his destination (Fig. 21-1). While this looked rather simple on his Jeppesen chart, despite great distances between nav aids (Boise to Kimberly, 163

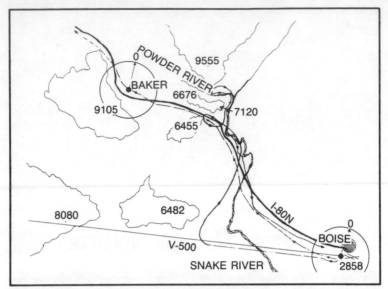

Fig. 21-1. The route flown by the pilot when he tried to get from Boise to Portland. The solid line is the first day's attempt. The dash-dot-dash line is the route flown on the following day.

nm, Kimberly to Newberg, 145 nm), the MEAs are pretty high along that route, 11,000 for the first portion and 10,000 thereafter. A study of the Sectional charts shows the terrain along the way to be mountainous and devoid of populated areas or places to land in an emergency. The reports and forecasts for the area involved talked of a frontal passage moving from the Pacific Northwest across the mountains southeasterly with frequent low ceilings and mountaintops obscured. But the visibility figures weren't too bad.

Though he was aware that with the reported weather conditions he might not be able to make the flight as planned, he took off, figuring that there is always a 50-50 chance that the reports were worse than the actual conditions. The climbout from Boise, though bumpy, was uneventful. But even after he had leveled off at 6,500 feet his ground speed was a depressing 95 knots based on the time it took to get to the point at which V-500 crosses the Snake River. Though the river itself and the terrain in the immediate vicinity is only about 2,500 feet or so, the mountains he would have to cross along his proposed route rise over 6,000 feet at first and eventually over 9,000 feet. By now he could clearly see that the weather straight ahead was hanging right down into the mountains (Fig. 21-2) and there was no question that he would have to detour and

146

try to find lower terrain if he was going to make it without having to climb into IFR conditions. The Jeppesen charts would be of no use in this attempt, so he spread the Sectionals on the seat to his right, annoyed by the fact that the only logical route available to him, namely following the Snake River as far as possible, lay on the edges and corners of three different charts, making chart reading in the bumpy airplane extremely difficult even though the autopilot was for the moment at least doing the actual flying of the aircraft. As best as he could figure out, he should be able to fly northward along the Snake to a point somewhere near a place called Huntington where a major highway turned away from the river toward Baker. He should be able then to follow that highway to Baker and from there to Island City and Pendleton where he'd intercept the Columbia River which would take him straight into Portland. According to the chart, all of this should be possible at under 5,000 feet without hitting anything.

At first everything went as planned. He flew north along the river, found the highway and followed it for a few miles, only to find that the pass between 6,000- and 7,000-foot mountains near a place called Durkee was totally obscured by low clouds and a horrendous rainshower (Fig. 21-3). He contacted the Baker FSS which reported a thunderstorm overhead with lightning all quadrants. Obviously, this wasn't going to work.

He made a fast 180 and flew back along the highway until he got back to the Snake River. But he still was not ready to give up. According to the chart, some miles north of his position the Powder River runs into the Snake from the west, and he figured that there might just be a chance that by following first the Snake and then the Powder River, he would be able to circumvent that pesky

Fig. 21-2. The weather ahead was hanging right down into the mountains.

Fig. 21-3. The pass ahead was obscured by low clouds and a horrendous rain shower.

thunderstorm. So, on he went, north along the Snake until he found the Powder where he turned left and followed it for a few miles in a westerly direction. Throughout all of this time the overcast above was intermittently broken, revealing towering buildups which appeared to reach right up into the jet altitudes, while the bases clung stubbornly to the tops of the higher mountains all around. It was kind of like flying in a tunnel, not a very comfortable experience, though the visibility itself remained ample below the clouds.

Well, the Powder River eventually fooled him too. The same thunderstorm, or maybe a close relative, blocked the way in no uncertain terms and there was just no alternative but to turn back. As much as he hated the idea of having wasted all this time, effort and fuel, he decided that he would rather be around to fly again some other day and he returned to Boise, checked back into the same Rodeway Inn and relaxed for the rest of the day.

The front passed through Boise that night, drenching it with several inches of rain, but by morning the reports were considerably more encouraging and he took off again. This time he made it, though the straight route along V-500 was still impossible VFR. But farther north, at Baker and beyond, the way he had attempted to go the day before, the clouds were scattered to occasionally broken and he was able to climb in VFR conditions to an altitude above the clouds. To his left, along the more straight line toward Portland, the tops still remained too high to be overflown without oxygen, but he eventually reached Pendleton, intercepted the Columbia River and followed it all the way to Portland.

The Nameless Creek

The second of our river adventures doesn't really involve a river at all, but rather an apparently nameless creek or rivulet which has its origin somewhere north of Toledo, Washington, and runs southward into the Columbia River. The flight was to start at Boeing Field in Seattle and was to terminate at Troutdale Airport in Portland, a total distance of only a little over 100 nm. The aircraft was a Bellanca Super Viking equipped with dual navcoms, ADF, DME, transponder and autopilot.

When it came time to leave Seattle, the weather was 1,500-feet overcast, visibility five, with intermittent rain showers and occasionally lower ceilings reported for the route between Seattle and Portland. The pilot was an experienced VFR pilot with several thousand hours in his logbook and some instrument experience but no instrument rating. After he and a non-pilot passenger took off from Seattle they leveled off at about 1,000 feet and flew across the southern portion of Puget Sound toward Olympia (Fig. 21-4). So far there was no serious problem, despite the low overcast. South of Olympia they followed the four-lane divided highway which runs more or less straight from there to Portland and remains at a fairly low elevation all the way except in the vicinity of Toledo, where the terrain rises a few hundred feet. There seemed to be no good reason why they shouldn't be able to make Portland without difficulty.

It was after passing the Chehalis-Centralia Airport that things began to get a bit sticky. No matter how optimistic they would have

Fig. 21-4. Level at 1,000 feet over Puget Sound.

149

liked to have been, there was no doubt that some miles ahead the highway and the base of the clouds were holding hands, a fact which was further emphasized by all those automobiles which were driving toward them with their headlights on.

Realizing that continuing on would be silly, they turned around and landed at Chehalis-Centralia to have a cup of coffee and figure out what to do next. While they were sitting at the counter in the coffee shop at the airport and discussing their predicament, a helpful local pilot told them about this little river which more or less parallels the highway and which usually remains flyable when the highway itself is obscured. They dragged out their charts and found that, true enough, if they followed the railroad tracks which ran right by the airport, they'd intercept something that looked on the chart like a tiny blue line, bordered on either side by a railroad and a road. Following it would take then eventually back to the major highway and to the bend in the Columbia River which they could then follow all the way to Portland (Fig. 21-5).

They decided to try it, figuring that, if it didn't work, they could always turn around again. They took off, picked up the railroad tracks and, after a while, found the little river. With the visibility below the clouds sufficient to permit them to be sure that there were no high-tension wires or other obstacles in the way, they stayed about 100 feet above the water, uncomfortably conscious that the terrain to both sides was obscured by clouds. But it worked. They made it to the Columbia River and then followed it around that great bend which separates Portland, Oregon from Vancouver, Washington, all the way to the Troutdale airport where they had to request a special VFR clearance for landing, because the ceiling was considerably less than 1,000 feet.

The Ohio River

There are certain similarities between this experience and the trip from Cincinnati to New York City, which we'll talk about now, but there were also great differences. It was September 1972, and the pilot had been at the National Business Aircraft Association convention in Cincinnati. When it was over he was asked by a friend if he would be available to ferry a Cessna 172 back to New York City. He was.

The distance from Cincinnati to New York City is about 550 nm which is well within the range of the Skyhawk, assuming flight at a reasonable altitude and with the mixture appropriately leaned. The original plan had been to fly from Cincinnati via V-128 to

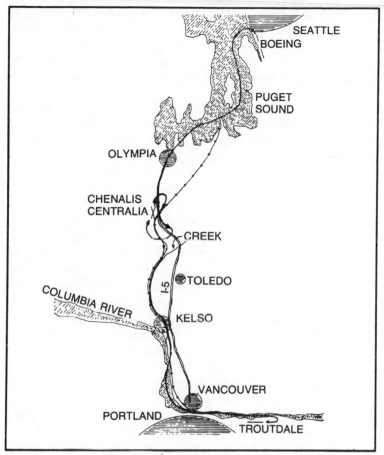

Fig. 21-5. Chart showing the flight from Seattle to Portland.

Charleston, West Virginia, and from there via V-4 to Elkins and Kessel, then take V-166 to Martinsburg, Westminister and New Castle, and then V-157 and subsequently V-123 via Robbinsville to Teterboro Airport. All pretty cut and dried if the weather had been good or if the pilot had been instrument rated. As it was, neither was the case.

The weather at Cincinnati on the day of departure was just barely VFR with a solid overcast around 3,000 feet and the visibility reported to be three miles, though it seemed considerably less. The pilot, with a student-pilot passenger in the right seat, took off at about 1,500 feet agl and headed south to pick up the 105-degree radial from the Cincinnati VOR (CVG) which coincides the V-128.

Once on that radial he changed course and flew outbound on it, hoping to be able to continue receiving the VOR until he would be able to pick up York (YRK), the next station along his route of flight, a distance of 84 nm. But it didn't work. The CDI needle began to shudder only too soon and then the OFF flag popped up and the needle went dead.

Now, except for the Ohio River which runs somewhere south of the airway, the country around there is rather featureless, with every mile looking like every other mile. He turned southward to pick up the river and use it as a guide, but by the time he got there the ceiling appeared to have lowered considerably, and he decided that it would be the better part of valor to land at some airport and to calmly figure out how to proceed from there. The airport he picked was Fleming-Mason, some short distance south of the river and, actually, only about 50 nm from where he had started out.

Taking the various appropriate charts with them, the pilot and his passenger settled down in the pilot lounge to assess the situation. When coming in for the landing at the airport, which is at a 915-foot elevation, the ceiling had been so low that it had actually been impossible to fly a regular pattern, and, as best as could be seen, to the east it looked just as bad, if not worse.

They spread the chart on a big table and discussed the next move. According to the long-range forecasts the weather was not expected to improve to any meaningful degree for at least a day, and neither was particularly enamored of the idea of spending all that time in a motel, even if there had been a motel, which there wasn't. The alternative was to find the best way to continue eastward. The terrain to the east rises to around 1,500 feet which, though certainly not high, was likely to be impossible considering the bases of the clouds in the area. But the Ohio River runs right through it at a much lower level, though its course is anything but straight. Still, they figured that they could take off, pick up the river and follow it along its convoluted route via Portsmouth and Ironton to Huntington, West Virginia. There the river makes a sharp turn to the north, but with the terrain east of there at a somewhat lower elevation, they might just be able to pick up the four-lane highway and follow it eastward to Charleston. From there it was likely to get a little complicated. Another river, the Kanawha, comes to there from the southeast out of Allegheny Mountains which they knew they couldn't possibly expect to cross unless there was a break in the overcast somewhere which would permit them to climb out to VFR conditions on top. The alternative was a rather convoluted highway which runs from Charleston along the western slopes of the Allegheny Mountains through Clarksburg and on into Pennsylvania. The longer they studied the charts, the more convinced they became that, while they could certainly make Charleston, there would be no alternative but to stay there and wait out the weather, unless they could obtain some information of improving conditions somewhere in the vicinity (Fig. 21-6).

Well, looking at the charts didn't improve things any, so they took off, picked up the river and followed it religiously around its many bends and turns. It was strenuous flying, and even at their altitude, just a few hundred feet above the water, the visibility was barely sufficient to give them an idea of the next bend in the river before it was time to make the turn. But at least they weren't in the clouds, though whether these conditions could have been described at legally VFR is doubtful. Instinctively both kept leaning forward as if that would improve their ability to detect powerlines or any other obstructions which might have to be avoided.

After what seemed like an inordinate amount of time they finally passed through Huntington. Seeing that the highway to their right appeared to be visible for at least a mile or so, they left the safety of the river and picked up the highway instead. By now, being within only about 35 nm from Charleston, the Charleston VOR (CRW) came in loud and clear and they were overjoyed to hear of occasional breaks in the overcast. If that held up, maybe they could make it after all.

They landed at Charleston, refueled and had a bite to eat. Then

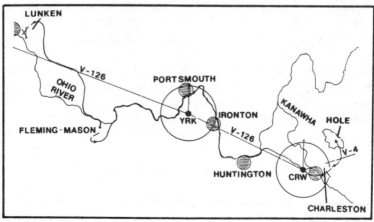

Fig. 21-6. Chart showing the flight east from Cincinnati to Charleston and on from there to VFR conditions on top.

they visited the FSS and found that, true enough, there were breaks reported in the overcast not too far east, and pilot reports about cloud tops placed them at between 10,000 and 12,000 feet, while the New York area was forecasting scattered to broken conditions with adequate visibilities.

The rest of the flight, then, turned out just fine. Though those reported breaks were not exactly big, fat holes, but rather thin spots through which the sun and sky above were barely visible, they did manage to climb out without being illegal for more than a minute or so, and once on top they could finally pick up a straight route toward their destination.

The Mississippi

The pilot who was sitting in Cedar Rapids, Iowa, and who finally ended up in Memphis, Tennessee, hadn't really wanted to go there at all. In fact, he was on his way home to Southern California, but one of those immense summer warm fronts had settled and then stalled over the entire middle of the country. And no change was forecast for some time. Being a pilot and, therefore, impatient, he decided that hanging around and waiting didn't appeal to him. Instead he would head south toward the Mississippi and follow it just as far as necessary to circumvent that warm front.

The Mississippi, of course, is a nice big river which makes flying along it easy, even if the clouds hang relatively low all around (Fig. 21-7). And this is what he did. The aircraft was a Piper Comanche 250 with 90 gallons of fuel, which gave him ample range. So he just flew on and on, often less than 100 feet above the water, always keeping an eye out to the west in order not to miss any chance which might permit him to turn in that direction. But it never worked. Luckily, as he came close to the St. Louis metropolitan area the ceiling did him a favor and lifted sufficiently to permit him to overfly all those bridges and other obstacles but after that it was down again, right onto the deck until he arrived some 35 or 40 nm north of Memphis (Fig. 21-8). Suddenly there was blue sky above and ahead, as if all that cloud deck had been cut with a knife. He went to Memphis to refuel and get a bit to eat, and then he could finally turn west toward Dallas and beyond, that awful warm front being now to the north of his course.

The James River

The pilot who started off from Charlotte, North Carolina, with

Fig. 21-7. The Mississippi is a nice, big river which makes flying along it easy.

Washington, D.C. as his destination never got there. And if it hadn't been for a bit of dumb luck, he might never have gotten anywhere at all. The total distance is less than 300 nm, and though he was flying a Cessna 150, which isn't known for its great range, he should have been able to make that easily. Except, of course, it didn't work that way.

The weather at Charlotte was amply VFR with a broken overcast at 5,000 feet and visibility better than 15 miles. But he hadn't been flying for more than 20 minutes or so when he decided that the turbulence below the clouds didn't appeal to him and that he'd rather be up above where it was bound to be smoother. So, picking a great, big blue spot between some of those broken clouds

Fig. 21-8. After that it was down again, right onto the deck.

he shoved the throttle all the way in and started up. He was right. Once he got above the clouds and was able to level out at 9,500 feet, it was smooth and simply beautiful. The forecast weather for the Washington area was much like where he was, broken conditions with good visibilities, so he wasn't worried.

He had been flying like that for about an hour when he realized that the undercast beneath him had become quite solid and that he hadn't had a glimpse of the ground for some time. He tuned his radio to Raleigh-Durham (RDU) and asked to be given the latest weather for Washington, expecting to be told that it was still as had been advertised earlier.

"Washington, ceiling 2,000 overcast, visibility three, light rain."

What had happened? Well, whatever had, if that continued he'd never be able to get down from his current altitude. He asked if there was a chance of improvement during the next few hours and was told that, to the contrary, the conditions were expected to deteriorate further. In that case, what was the nearest place with reports of broken conditions or better? He was asked to stand by, and after a pause which seemed much longer than it actually was, the radio came back to life with the intelligence that everything north of here was either actually IFR or at least covered by a solid overcast, with only Lynchburg, Virginia, reporting occasional breaks in the overcast.

Lynchburg? Where is Lynchburg? He dug out his charts and eventually found it on his Cincinnati Sectional. So that was Lynchburg, right next to some 4,000-foot mountains and with the

Fig. 21-9. He searched for the breaks in the clouds but couldn't find any.

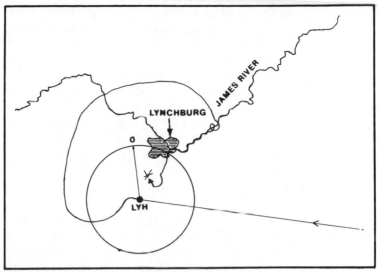

Fig. 21-10. Chart showing the flight to Lynchburg.

James River running right through it. Well, Lynchburg it was going to be. He tuned his nav receiver to 109.2 and was gratified to have his OBI react immediately. After all, even though he still had about half his fuel, he didn't relish the idea of running low while looking for some of those occasional breaks which, he hoped would have the decency to stick around until he got there.

As he had many times before, he was wishing he was flying a faster airplane. Even though the distance from his present position to Lynchburg wasn't that great, it just seemed to take forever. But finally, finally the CDI started to fluctuate and then the ambiguity meter changed from TO to FROM, meaning that he had just overflown the VOR. But what about those breaks? He searched the undercast in all directions but could see nothing but solid clouds (Fig. 21-9). He called Flight Service while listening through the Lynchburg VOR and asked for the latest report and was told exactly what he'd been told a while earlier: Fifteen hundred feet overcast, visibility six, occasional breaks in the overcast.

Okay, so all he had to do is find those breaks. He decided to fly a wide circle around the VOR, turning first south, then west, then north and so on, and, true enough, there, off to his right, was something that might just turn out to be what he had been looking for. He headed for it and when he got there what he saw was a tree-lined river right below his wing. Without further ado he pulled back on the throttle and put the nose down toward that friendly piece of real estate. He had to do a bit of fancy twisting and turning in order to stay out of the clouds, but pretty soon he was down there over the river with the bases of the clouds hanging low all around. He knew, of course, that by following the river he would eventually get to Lynchburg, except he didn't know which way to follow the river. He could have kicked himself. Why hadn't he checked his VOR receiver before heading down so that he would now know in which direction the station was from his present position? As it was, way down here in the riverbend between the hills, the VOR stubbornly refused to come in again. And, even though he was fairly certain that he was somewhere north of the city and the airport, that damn river was twisting and turning in all kinds of directions, and it did, in fact, turn north on both sides of the town. Well, eenie, meenie, minie, moe . . . and he turned in the direction which seemed to him, for some reason, to be more logical than the other.

This is where his dumb luck came in. As it turned out, he had picked the right direction (Fig. 21-10). After about five or 10 minutes of carefully following every turn in the river, there were some houses, and then some more, and even the OBI came back to life and after a bit of low-level searching he found the airport and was glad to finally be on the ground again, even though it wasn't where he had wanted to go.

Chapter 22

Tailwind I

In later years the flight was to be one to talk about time and again in pilot lounges and drafty hangars. After all, how many pilots could claim to have covered 482 nm in two hours and 10 minutes in an airplane that had a normal cruise speed of 100 knots? It sounded more like a fairytale but it did, in fact, happen.

It all started at El Paso on a windy spring day. The pilot, after preflighting his Cessna 175, removed the tiedown chains and started to taxi toward the active runway, firmly holding onto the yoke to prevent the rudder and ailerons from being buffeted by the strong gusty wind. Then came the first of several unexpected developments that day. At an intersection of several taxiways he tried to make a 90-degree turn downwind, but the airplane simply wouldn't respond. No matter how hard he tried, it simply wouldn't turn the way he wanted it to. At first he couldn't figure out the reason, and then it occurred to him that the wind was pushing so hard against the vertical stabilizer that it made a turn in that direction impossible. Okay, if it wouldn't turn one way, maybe it would turn the other. So, instead of making a 90-degree turn to the right, he tried for a 270-degree to the left and, sure enough, the desired result was achieved.

At the takeoff end of the two-mile-long runway he did his runup, checked the controls, and called the tower which cleared him for takeoff. With the wind right on his nose he seemed to be lifting off before he really had a chance to get moving, but the airplane was safely airborne and climbing the way it should. His takeoff had been on Runway 26 so now, with his destination being Waco, Texas, and the planned fuel stop Midland, he started his turn toward the east while, at the same time, calling the FSS to ask them to activate his VFR flight plan. Halfway through the turn he looked at the ground and was aware of being blown sideways across the ground, but eventually he was lined up and on course toward the east, cussing under his breath because he was getting bounced around something fierce.

Getting Above the Turbulence

Though he had accumulated several hundred hours in the airplane, he had not yet reached the stage where he would accept uncomfortable turbulence with equanimity. Instead, whenever things got rough, he would climb to higher and higher altitudes under the assumption that things might get smoother up there. Most of the time this does prove to be the case, but on this day it didn't seem to work. He was already climbing through 10,000 feet and still the turbulence refused to abate. There were some scattered to occasionally broken clouds just above him and he decided to try and get on top of them, hoping that that might do the trick. Though he had no oxygen on board, he had often before spent considerable time at 14,000 or 15,000 feet without any more ill effect than a slight headache.

At 13,500 feet he was finally on top of the clouds, for all the good that climbing had done him. It was still rough, so he knew he'd just have to try and relax and enjoy it. Obviously, this was going to be just one of those days.

He has opened his flight plan at 10:30 a.m. Mountain Standard Time or to be more professional, at 1730 Zulu and had figured his time en route to Midland as two hours and 10 minutes. The climb to his present altitude had taken 33 minutes at an airspeed considerably below his normal cruise, so he would probably be late and as soon as he could reach Midland VOR he would call them down there and tell them to extend his flight plan by some 15 or 20 minutes. Well, there was plenty of time. He was certainly still way beyond reception distance.

He was still receiving the El Paso VOR and was navigating outbound on the appropriate radial which he knew would take him via Salt Flat and Wink to Midland. Looking down at the West Texas landscape which offers few landmarks other than the Guadalupe Mountains just east of Salt Flat, he was surprised to notice that the mountains were not where he had expected them to be. But then after all, he could only see portions of the ground through the spaces between the clouds, so he was most probably mistaken. Still, just to make sure he tuned in the Salt Flat VOR and when the ambiguity meter in the OBI produced a FROM rather than TO reading, he briefly thought that maybe there was something wrong with the nav receiver. He played around with it for a while, tuning in other VORs in the area, Hudspeth and Carlsbad, only to find that either something was completely crazy, or he was much farther along than could reasonably have been expected. Well, he might as well try Midland and see if he could receive them. Sure enough, not only did Midland come in loud and clear, but if his OBI wasn't lying to him, he was practically there already.

Amazing Journey

How could that be? By now he'd been airborne for just over an hour, and the distance from El Paso to Midland was, after all, over 200 nm. How could he have covered that much ground in so short a time in an airplane that had never before managed to cruise at better than 100 knots? Maybe it wasn't possible, but there was a big opening between the clouds and just below him and to the left he couldn't fail to recognize the Midland airport. He'd been there often enough before to know exactly what it looked like. Okay, so maybe he was flying a jet and didn't know it. Anyway, he called Midland and told them to cancel his VFR flight plan, saying that at his apparent current ground speed he had ample fuel on board to continue on non-stop to Waco.

Throughout the rest of the flight he was so busy marveling at the unbelievable story which his nav receiver insisted on telling him as he tuned from one VOR to the next, that he practically forgot to be annoyed at the ride which never had abated. Exactly two hours and 10 minutes after liftoff his wheels touched down on Runway 23 at Waco, Texas.

Still somewhat disbelieving about what had happened during the past two hours, he went up to the tower to see if those guys up there might have a solution to the puzzle.

"Hi."

"Hi. Listen, did you guys move Waco from where it used to be?" he asked.

"Not that I know of, why?"

"I just came here from El Paso in a Cessna 175 which cruises at 100 knots and it took me two hours and 10 minutes. It just doesn't make sense."

"How high did you fly?"

"Pretty high. It was pretty bumpy so I climbed up to 13,500 and stayed up there all the way."

"That explains it."

"How come?"

"We've got a freak jetstream situation today. It's way down low."

"You mean, I've been flying in the jetstream?"

"Seems that way. We've had reports of winds up to 150 knots as low as 15,000 feet."

"Heavens! It's darn lucky I wasn't going the other way. I'd have been flying backwards."

This then was the tailwind which, for years to come, would make up for hours and hours of headwinds, and which was to become the subject of much hangar talk.

Chapter 23

Tailwind II

When the pilot awoke that morning he was discouraged by the lack of direct sunlight coming through the window because that probably meant there was a heavy overcast outside. Sure enough clouds hung overhead with a stiff north wind blowing, but thankfully there was no rain or other precipitation that could hinder visibility.

Naturally, he had to be back in the office the next day for some important work and he wondered if he should return today or wait one day. He called Flight Service to see if the clouds were local, hoping they would dissipate by the estimated time of departure that afternoon.

The forecast wasn't good but it wasn't bleak either. Clouds covered the area for several hundred miles around, including his direction of flight, and they were not expected to break up for at least 24 hours. However, it was only an overcast with no rain or fog or other hinderance to visibility that, in fact, ranged from 10-20 miles along the route. The wind was 20-30 knots from the north and he was flying south so there would at least be a tailwind.

If the forecast held it would be a legal VFR flight but it would be at a very low altitude. Fortunately, it was a very familiar route for the pilot—learning to fly over the first half of the route and covering the second half twice a month on business. Highways, cities, lakes, powerlines, and other landmarks were memorized along with the most important element—airport locations. But it was a familiar route from several thousand feet at cruise altitudes. How would it be at traffic pattern altitudes? Go-no go? Probably go, but wait until early afternoon and check the weather again . . . .

Getting Apprehensive

By lunchtime, the local weather had not changed and the pilot was getting slightly apprehensive. Flight Service reported some breaks in the clouds to the north, but along the route of flight conditions were unchanged—good visibility with a tailwind. Based on that information the pilot filed a VFR flight plan and started preparing to leave.

Complementing his knowledge of the route, he took time to closely examine the Sectional chart targeting airports that could become safe refuge if conditions worsened. Fuel would be no problem, so if the weather held it would be a non-stop flight. Go-no go? By now it was "go"—to the airport—and he was on the way to the airport.

His apprehension was allayed en route to the airport with sunlight punching through the overcast here and there. Airspace illuminated by the sun below the clouds was crystal clear—no moisture to hinder visibility. Buffeted by the north wind he loaded the plane, preflighted, and went inside to check the weather one last time before taking off. Flight Service reported no change for better or worse—"go".

Airborne, he took up a heading that would lead him to the route's major dogleg and he climbed to find the ceiling high enough to cruise nearly 1,000 feet above ground level (AGL). Leveled out, he took stock of the situation—aircraft performing well, good visibility, some minor bumps, and what appeared to be a great ground speed—and relaxed. For the moment, everything was okay and if it deteriorated rapidly he could simply turn back and try again the next day.

Primary checkpoints came up quickly and were right on target with little course correction, so he was riding a direct tailwind. His "lifeline" for the first leg of the flight appeared quickly and he relaxed some more. The major four-lane highway would take him directly to the dogleg turn and he was right on top of it. His earlier apprehension gave way to enjoying some unseen sights when flying the route at higher altitudes. Then the navigation system came to life receiving the VOR representing the dogleg's joint.

Rapid Changes

The pilot was again taking stock of the situation when he

realized it was getting darker. Obviously the overcast was getting thicker but he could still see more than 10 miles underneath so he was not worried. Then he realized how the clouds were getting thicker—they were getting lower. He had to drop down several hundred feet to avoid the clouds. Thankfully the ceiling leveled out and forward visibility was holding so the clouds did not drop to the surface.

Approaching the VOR, it was time for an en route go-no go? decision. The flight was progressing better than expected; thanks to the tailwind it was at least 15 minutes ahead of schedule compared to other trips. The aircraft was performing flawlessly. Visibility was still excellent. Fuel was fine. The only worry was the cloud cover and the question of whether or not it would hold or drop lower.

A quick visit with Flight Service revealed good visibility at the destination, no rain, or fog reported along the remainder of the route and it appeared the ceiling did not change much. The pilot's first en route decision was "keep going".

Negotiating the dogleg would be tricky. He wanted to approach the VOR and establish his heading FROM the station before arriving at the station. But a busy airport near the VOR would prevent that. He flew TO the station at his present heading and made the righthand dogleg turn just prior to arriving over the VOR site that he could see. All of a sudden, he realized it had gotten darker again and he was gradually descending to avoid the cloud deck, but he got on his radial and verified it with references to his compass and directional gyro. He waited for a new road to appear.

He knew at such a low altitude that VOR reception would be lost soon and he had to be over the road that would keep him on course. Here came a road and comparing it to the Sectional it appeared to be the road he needed so he made the course correction and started following the road. He was beginning to get apprehensive again. He had driven on the road he wanted to follow and this road had no familiar landmarks like houses or gas stations.

He decided to hold this heading and see where the road went, hoping it would get familiar soon. Turning back never crossed his mind at this point. He was ahead of schedule and he wanted to be home, not halfway. However, it would get worse before it got better.

He was over the wrong road and by now it had twisted and curved until he was heading north, 90 degrees perpendicular to his intended route. He quickly decided to make a 180 degree turn to intersect the road he wanted to follow. For the moment, he was "disoriented" but he could always take up a heading that would go back to the VOR's reception area and start from there again, so the pilot continued searching for his east-west highway.

Looking For a Lake

Flying due south now he was over a north-south road that hopefully would intersect his highway. According to the Sectional, a lake would be visible south of the roadway's intersection, so that made for a bonafide checkpoint. He continued looking but there was no lake or any major intersections . . . then his north-south road stopped. Apprehension finally gave way to real fear . . . he was lost. He flew a crude figure eight hoping to spot a water tower, town, or, best of all, an airport, but there was nothing . . . just pasture land. Attempts to raise any VOR was futile; he was too low.

His first decision was the smartest he could make . . . *Think* . . . take up a straight and level course that will at least head in the correct direction toward home . . . settle down and think some more. He needed someplace to fly, a specific place that would ideally get him back on course or at least have an airport where he could land and take stock of the situation. Worst case scenario would be to climb and hope the overcast was thin enough to break out on top and take it from there.

He scanned the horizon but it remained barren. He took a deep breath and decided to examine the Sectional. Considering the major course corrections he made earlier—wrong course corrections—he mentally drew a box where he was probably located. Scanning the chart he looked for "help" in the area around the box. He spotted something . . . maybe it would work.

A small airport he had flown over many times before had a nondirectional beacon (NDB) that just might be strong enough to receive . . . wherever he was. Quickly tuning the automatic direction finder (ADF) the audible morse code identifier filled the cockpit and the needle swung hard locking onto the signal. Relief— someplace to fly, and it would get the trip right back on course!

Established on course to the NDB the pilot did a fast analysis and found he was far south of this intended track that would have been to the west. After the VOR dogleg he apparently followed a highway that turned north instead of running west. Then his road that ran south was not the road near the lake—he had flown past the intended east-west highway while looking for the lake.

Back on Course

Approaching the NDB and spotting the airport, he wanted to

land and take a breather, but the weather was holding, fuel and aircraft were fine, and he was starting to settle down—he knew exactly where he was and he knew exactly how to get home. Also, three airports would be right on course and a fourth off-course airport would be visible if needed. Still "go".

Like clockwork the flight continued. The ceiling even seemed to lift some and the pilot tuned the ADF to receive the AM broadcast station at home—it came in loud and clear. He would remain over major highways until about 30 miles away from the destination airport, then it would be up to the airplane's ADF to point the way home. Two of the three airports passed by and the airplane approached the third. Time for another tough decision—land at the third airport and call for a ride home or continue? Critical aspects of the flight were considered—visibility was excellent, the plane had plenty of fuel, the ceiling had not dropped any lower, there was no turbulence, the ADF was working flawlessly. Directly over the last airport he made the decision—continue. He had no suspicion of what was about to sneak up on him.

The Microwave Tower

Relying solely on the ADF and the broadcast station he was right on course for home. The pilot glanced over his shoulder and looked at the last airport as it disappeared. When he turned around the cloud deck seemed to drop all of a sudden. He descended slightly and looked forward to be sure the good visibility was holding. It was fine, but the cloud to ground distance seemed to be deteriorating. Were the clouds lowering or was the terrain rising?

The aircraft was too low for a quick turn back to the last airport and by now the ADF was receiving the home airport's NDB, so the flight continued. The terrain had risen, but it leveled out and familiar roads fell into place under the aircraft. The pilot knew he was extremely low, rarely glancing at the altimeter, more concerned with staying away from the ground—and a microwave tower in the area.

He knew the tower was on the left side of the road he was flying over. And the ADF was pointing left of course to the airport, so he made a 45 degree left turn, heading for the airport and watching for the tower. Thankfully it appeared on the right side and the plane passed by at least a mile away.

Now the only thing lying between the aircraft and the airport was the hometown. Ideally, the pilot would skirt the city, approaching the airport from the southeast or northeast, but there were power transmission lines south of the airport and radio towers on the north. He would have to fly very low over the southern part of the city. There were no major obstacles but he was sensitive to noise created by the airplane, so he reduced power for a quiet overflight. And by flying over that section of town, he could safely glide to one of several pastures in an emergency, away from homes and businesses.

Finally, the uncontrolled airport appeared with the windsock indicating the same stiff north wind that was present at departure. The pilot crossed over the airport at mid-field and entered the prescribed right downwind that kept traffic away from the city. Abeam the touchdown zone he reduced power, kept his landing pattern close since he was already so low, crossed the numbers and touched down—safely home.

Revelations

He taxiied up to the fuel pump, pulled the mixture back, and breathed a tremendous sigh of relief as the prop jerked to a stop. A quick call to Flight Service cancelled his flight plan. Only then did the pilot realize what time it was. Usually, the trip averaged two-and-a-half hours. This trip from wheels off to wheels down had taken just two hours. A remarkable trip, but imagine how fast it would have been without the deviations . . .. Closing the hangar doors the pilot completely relaxed as he latched the padlock.

He looked up at the clouds for the first time since landing. He paused to think . . . if he were *opening* the hangar instead of *closing* it, would he fly in the prevailing weather conditions? Probably not.

Chapter 24

Accident?

The pilot had taken off from Louisville, Kentucky, at 10 p.m. one summer evening, on a flight to Lansing, Michigan. He was a private pilot with 300-plus hours in the left seat. He was alone, flying a Skyhawk which within the past 10 hours had undergone an annual inspection. The weather briefing he had obtained prior to takeoff had indicated ample VFR conditions all along his route of flight, the only reported weather being a line of thunderstorms moving toward Michigan from the northwest. There was more than adequate fuel on board to make the trip non-stop even if, as the weather reports seemed to indicate, he would have to contend with some moderate headwinds.

His planned route of flight would take him over Muncie and Fort Wayne, Indiana, to the Litchfield VOR and from there to Lansing, a total distance of 285 nm. Even though he had not filed a flight plan, he contacted the various FSSs along his route to report his position and to obtain information about any possible changes in the weather along his route and his destination.

The flight proceeded without difficulty despite the high overcast which obscured whatever light might have been available from the moon or stars. The visibility proved to be excellent all along the way and he congratulated himself on the precision of his navigation which caused him to hit every VOR right on the nose. Even his estimated time between nav aids turned out to be correct within a minute or so on each leg. At 1:48 a.m. he overflew the Litchfield

VOR and some seven or eight minutes later he contacted Lansing Tower, reporting his position as 20 miles south of the airport.

"Lansing Tower, Cessna Three Two Five Niner, two zero south at 2,500, landing Lansing, over."

"Cessna Two Five Niner, continue approach."

"Two Five Niner."

"Lansing Tower, Learjet Five Sierra Foxtrot, ready."

"Five Sierra Foxtrot, cleared for takeoff."

"Rolling."

The pilot wondered if he would ever be able to learn to affect that bored and tired sound in his voice that seemed to be a sort of trademark of airline and professional pilots. Somehow you could always differentiate between the old timers and guys like himself by the way they sounded on the radio.

There were broken clouds below him, but he could clearly see the lights of the city through the wide open spaces between the clouds. He let down and after a while was able to pick up the rotating beacon of the airport on the far side of the city proper.

"Lansing Tower, Cessna Two Five Niner, how about a straight-in approach?"

"Cessna Two Five Niner, roger, make straight-in to Runway 32, wind 290 degrees at niner. Report two mile final."

"Two Five Niner, roger."

Request Your Intentions

He wondered how he was supposed to know when he was two miles out, but he'd take a guess at it as he had always done. By now he was below the broken overcast and had the runway lights clearly in sight. He was just about to pick up the microphone to tell the tower that he estimated his position as being on two mile final, when they called him.

"Cessna, Two Five Niner, What's your position now? Over."

Rats. The mike had slipped out of his hand and dropped to the floor.

"Cessna Two Five Niner, Lansing. Do you read?"

He pulled the mike back up by its cord and pressed the button. "Ah . . . Two Five Niner."

"Roger, Two Five Niner, if you read now, the field is IFR, measured ceiling 900 broken, 3,000 overcast, visibility 10. Over."

"Two Five Niner." *So what? The airport is right there.*

"And, Two Five Niner, request your intentions, over."

What's he mean by that? "Say again?"

"Roger, the field is IFR, request your intentions, over."
What does he want from me? "I'm not reading you."
"Roger. The field is IFR with a measured ceiling 900 broken. Request your intentions, over."

I'll Go On Over

In the tower cab the controller watches the landing light of the approaching Cessna above the lights of the city below. Just as he was about to ask once more what the Cessna pilot wanted to do under the changed circumstances, his voice came over the speaker.
"I'll go on over."
"You'll be overflying the airport. Is that correct?"
"Right."
"Roger."
The controller picked up the direct line to the Flight-Service station and asked if there was a flight plan for an inbound Cessna Three Three Two Five Niner.
"Uh uh, haven't got a thing inbound, not a thing."
"Okay, thanks. He just called for a landing, then said he's going to overfly the airport." He hung up the phone and noticed the rotating beacon coming to life at the ramp.
"Lansing. Triple Six Quebec."
"Triple Six Quebec, go ahead."
"We're going to Minneapolis; you got our clearance?"
"Westwind Triple Six Quebec cleared to Minneapolis-St. Paul International Airport via Victor-2, Muskegon, Victor-450, Green Bay, Victor-26, Eau Claire, Victor-78, Minneapolis. Departure-control frequency 125.9."
"Two Muskegon, 450 Greenbay, 26 Eau Claire, 78 Minneapolis. One two five point niner."
"That is correct."
"Ready to taxi."
"Taxi Runway 27 or 32 your choice. Wind 330 degrees at niner. Altimeter 29.35."
"Okay. Taking two seven."
He watched the lights of the twin jet as it started to taxi, then looked around for a sign of the Cessna, but couldn't find it.
"Lansing, Triple Six Quebec is ready to go when we get to the end of the runway."
"Roger, cleared for takeoff."
"Okay."

"Lansing Tower, Cessna Three Three Two Five Niner, do you read?"

"Two Five Niner, make right turn to the north. Advise you that the field is IFR and traffic departing Runway 27, over." *Where is he, anyway?* "Westwind Triple Six Quebec, if you read, hold in position."

"Okay."

"Cessna Two Five Niner, Lansing. Do you read?"

"I read you very well, Two Five Niner."

"Roger. The field is IFR. Turn to the north, fly to the north. Traffic departing Runway 27 going west-bound."

"Two Five Niner."

"Two Five Niner report when you're clear of the control zone. Traffic is waiting to depart."

"Two Five Niner."

He waited a while, then called again. "Two Five Niner, I do not have you in sight now. How far are you, over."

"Ah . . . fifteen miles."

"You say, 15 miles?"

"Affirmative."

"Roger, continue."

"Lansing, Westwind. What's it look like?"

"Westwind Triple Six Quebec, cleared for takeoff and the Cessna says 15 miles out, but I'm not sure. He's supposed to be north."

"What altitude is he?"

"He appeared to be about 900. He's VFR. Cessna Two Five Niner, what is your altitude?"

"Two thousand five hundred."

"Roger, 2,500."

At this point the light on the direct line to the weather bureau lit up and he picked up the receiver. "Go ahead, weather."

"Yeah, if you see any lightning, let me know. The radar indicates increasing activity all through the area here, but I don't see any lightning."

"No, I haven't seen any yet, either. I heard Center talking to Muskegon and they said they had a big line over that way."

"Yeah, there's some stuff, ah, 42,000 feet, over Grand Rapids direction, but I don't see it, don't see any lightning here."

"Well, I hadn't been watching so far. Didn't pay much attention. Will let you know when I see something."

"Okay, fine. Thank you."

"Westwind Triple Six Quebec, contact departure."

"Triple Six Quebec, good night, now."

"Good night." He looked around the sky in all directions, but there was nothing to see. "Cessna Two Five Niner, Lansing tower."

"Lansing tower?"

"Roger, Two Five Niner. Lansing weather is now measured ceiling 900 broken, 3,800 overcast. The field is IFR and, ah, request your intentions, over."

"I'm going over to Grand Rapids."

"Roger, you're proceeding to Grand Rapids, is that correct?"

"Affirmative."

Cessna Three Three Two Five Nine never landed at Grand Rapids. The wreckage and the fatally injured pilot were found the next morning in a field, some six miles west-southwest of Grand Rapids airport, having apparently impacted the ground in a near-vertical attitude.

The Special VFR Approach

It is hard to believe that the sequence of events recounted above is based on the transcript of an actual case. The location and identifications were changed for the purpose of this dramatization, but all pertinent radio transmissions actually did take place. Our reason for incorporating this unhappy occurrence here is to demonstrate how a controller, by performing his duty exactly by the book, can, in fact, become the cause of an accident. After all, the Cessna was established on final approach and had the runway in sight. There was no reason on earth for him not to continue the approach and land, except that the airport suddenly turned technically IFR and thus, according to the rules, the pilot would either have had to request an instrument approach for which he was not licensed and which might have been beyond his capabilities, or he could have requested a special VFR approach. Obviously he was unaware of this alternative, or the sudden and unexpected turn of events confused him so that he simply forgot.

Despite the fact that controllers, under normal conditions are not authorized to suggest a special VFR clearance to a pilot, this was clearly not a case of normal circumstances. He could even have used a phrase such as: "Do you intend to request a special VFR approach," which would have kept him legal in that he would not actually have suggested such an approach. Instead, by asking the pilot his intentions over and over again, he, most probably, simply added to the pilot's confusion.

And then, on top of everything else, when the pilot told him that he was going on to Grand Rapids, he did not inform him of the severe weather in that area, even though, just minutes before, the weather bureau had mentioned buildups of 42,000 feet in that area.

An experienced VFR pilot, finding himself in this predicament, would either have simply pretended that he didn't hear the tower and would have landed, or he would have used the special VFR available to him. When a safe landing is clearly assured, it makes absolutely no sense to let oneself be diverted into a situation which, as in this case, can prove to be dangerous if not fatal.

Chapter 25

Loaded

The pilot was about to embark on a lengthy trip in his Tripacer, starting at Santa Monica, California, and eventually taking him to the East Coast. With him were his wife and his son's girlfriend; they were going to meet his son in Boston where he was going to college. Ignorant of the importance of weight when traveling by light aircraft, the two women had loaded the car with every conceivable piece of luggage and the pilot himself was also carrying quite a bit of stuff, because he planned to do some business while in the East.

The weather, on the day of the planned departure, was beautiful with none of the low clouds, fog and smog which so often plague the Los Angeles Basin. Once at the airport the pilot made sure that the Tripacer was fully fueled and then proceeded to load all that luggage into the back of the airplane where it filled not only the luggage compartment but was also piled high on one half of the back seat.

Suddenly The Airplane Tipped

The pilot had not yet accumulated a great amount of experience. He had obtained his private license only months before and, after purchasing the second-hand Tripacer, had only made a few trips in it, always alone. His flight training had failed to include any lessons on the subject of weight and balance, and it never really occurred to him that with three grownups and all that luggage aboard, he might find that the airplane would balk. The first sign that something was awry occurred when his son's girlfriend climbed into the backseat before he himself had strapped himself in. Suddenly the airplane tipped back, raising the nosewheel off the ground and resting on its tail. But the pilot still wasn't alarmed. He figured that as soon as his weight would be in the front seat, the situation would automatically rectify itself. And, for the moment, at least, it turned out that he was quite right. As soon as he and his wife were settled in the front, the plane righted itself and there seemed to be no further problem.

He taxied to the end of the runway. Since it was one of those rare days with an easterly Santa Ana wind, Runway 3 was in use. He would be taking off to the east over the city of Santa Monica and would be able to continue straight out over Los Angeles and on toward the Beaumont Pass. Cleared for takeoff, he pulled into position, fed in full throttle, and was surprised to find that it seemed to take a lot longer than usual to become airborne. But eventually the stubby little airplane did lift off and started to climb at a depressingly slow rate. Granted, he attained sufficient altitude to safely clear any obstacle, but all he seemed to be able to coax out of the airplane was something less than 100 fpm. In addition he slowly became aware that, even though the nose was trimmed down, the airplane insisted on flying in a nose-high attitude, mushing rather than flying the way it had usually done.

By the time they were passing over the eastern portion of Los Angeles they had still only managed 1,600, or 1,700 feet and though he had never before flown out of Los Angeles to the east before, he knew from his charts that they'd have to get up quite a bit in order to clear the various mountains between here and Phoenix. He was only then beginning to realize that the problem was probably not with the airplane but rather with all that extra weight. Something had to be done because they couldn't possibly fly all that planned trip with the airplane behaving the way it did.

Time To Unload

Knowing that both his passengers were in a light airplane for the first time, he hated to add to their quite natural apprehensions by admitting that something was wrong, but there just wasn't any alternative. So he told them that he'd have to go back to Santa Monica and get rid of some of the luggage because the airplane was quite obviously overloaded.

By now he was actually beginning to worry about keeping the plane in the air and he made a very wide and shallow 180 and headed back. It worked all right. They got back to Santa Monica, landed, and to the dismay of the two females, stored every superfluous piece of luggage in the car, locked it, and then tried again. This time the Tripacer behaved the way it should and, only a little over an hour late, they were again on their way to Phoenix, El Paso and beyond.

Later on, after the pilot had accumulated a fair number of hours in the left seat of a variety of airplanes, thinking back on this day made his skin crawl. By now he had become familiar, in theory at least, with the drastic effects which can result from ignoring the weight-and-balance limitations of an airplane. Especially with too much weight concentrated aft of the center of gravity, it is quite possible to set up a situation in which the aircraft can no longer be trimmed for level flight and despite the best efforts of the pilot, it may actually stall. On that particular day it was a lucky break that the air was smooth because if there had been any serious turbulence, it could easily have resulted briefly in an angle of attack too extreme to keep the airplane from stalling. And a stall with that load condition would, more likely than not, have turned into a spin from which recovery at that altitude would have been impossible.

Chapter 26

Panic

It was late afternoon. Dark clouds rimmed by the golden light of the lowering sun were gathering overhead. A storm which had been building for days over the Pacific south of Alaska was creeping inland across the Siskyou Mountains and Trinity Alps of northern California, spilling its clouds between tall peaks toward Red Bluff and into the valley which lies protected between the Sierra Nevada and the coastal mountain range.

The forecast for the area called for ceilings of less than 1,000 feet and visibilities of less than three miles with light to moderate rain showers. Conditions were expected to continue throughout the evening and into the night.

Looking up, there were still specks of blue sky, blue with the orange fringe of sunset between the billowing cumulus (Fig. 26-1). For a brief moment, the golden shape of a sunlit Mooney moved across, disappearing again behind the fast moving clouds. From its vantage point the sky was a fairyland of deceiving beauty. Colors ranged from aquamarine to deep purple. In the west, where mountains topping the clouds tried to reach for the heavens, the golden sun played hide and seek. But the pilot of the Mooney was not in the mood to contemplate beauty. Caught on top, he found himself pulling his airplane somewhat erratically around the towering buildups, his eyes desperately searching for a glimpse of the ground, a hole in the undercast large enough to permit him to get down. Racing with what seemed to him to be incredible speed

Fig. 26-1. Looking up there were still specks of blue sky.

between gold-white-purple-black ever-moving shapes which threatened to swallow him and his airplane, he began to realize that he no longer had any clear idea of his present position.

"I wonder how high the mountains are around here." Like so many pilots who often fly alone, he tended to do his thinking out loud.

"How high am I? Over 12,000 feet. That's good. There couldn't be anything that high." He looked at the gauges. "At least there's still enough fuel. Maybe not an awful lot, but enough to get me down."

His radio, tuned to the Red Bluff VOR, suddenly came to life. "Roger, seven three Tango. Cancelling your VFR flight plan. Latest Red Bluff weather 1,800 broken, 2,500 overcast, visibility three, light rain, wind light and variable, altimeter 28.83. Do you have the airport in sight?" Pause. "There is no reported traffic."

The pilot breathed a sigh of relief. "That's the place to go. They're VFR and it's got to be down there someplace." He centered the CDI needle, obtaining a 243-degree reading and a TO indication. Suddenly it was dark in the cockpit. He looked up. "Shucks." He was in the clouds. His hands automatically tightened on the yoke as turbulence bounced him first up, then down. "I wish I had an autopilot. What's that everybody keeps saying? Never mind the seat of the pants, just watch the instruments—airspeed—artificial horizon! How come it's slowing down? Ninety-five, ninety . . . Push that nose down! Ahhh . . ." Suddenly there was light again in the cockpit. He was once more in the clear. The sky above still glowed

Fig. 26-2. The sky above glowed with the evening light.

with the evening light, a star appearing here and there (Fig. 26-2). But all around, right, left, in front and behind, as far as he could see, immense buildups were reaching toward the heavens.

I've Got To Get Down

He banked, circling to try and stay in the clear and with no horizon to guide him he instinctively increased backpressure, as if, by climbing steeper and steeper, he could outmaneuver the clouds (Fig. 26-3). A stall warning beeped briefly and erratically he pushed the yoke forward. "I've got to get down through this

Fig. 26-3. With no horizon to guide him, he instinctively increased back pressure.

stuff." He tried to remember what had been said about the ceiling at Red Bluff, but he couldn't. He picked up the mike.

"Red Bluff Radio, Mooney Three Four Two Whiskey Alpha, over."

"Two Whiskey Alpha, Red Bluff, go ahead."

"What's your weather down there?"

"Stand by."

Why don't these people simply look out of the window? Stand by; I can't stand by!

"Two Whiskey Alpha, Red Bluff reporting 1,200 scattered, 1,800 broken, 2,400 overcast, light rain. Visibility three."

"Roger."

"Two Whiskey Alpha, what is your present position?"

The heck with my present position. None of your business! He tried to hang the mike back on its hook but failed as turbulence bounced him around and he simply hung it over the right yoke.

"Two Whiskey Alpha, do you read Red Bluff?"

Forget it! I'm not going to have the FAA jump all over me for coming down through this stuff without an instrument ticket.

Right then he was moving toward an immense cloud straight ahead of him, which, like some undulating prehistoric monster, blocked his way. With all the self control he could muster he concentrated on the airspeed indicator and artificial horizon, trying to ignore the increasing inner tension as all visual contact with the outside world disappeared. He quickly glanced at the OBI. *What happened?* The needle had pegged. Hesitantly he took one hand off the yoke and turned the OBS knob. The needle flipped to the other side, then slowly centered. The OBI indicated 183 degrees, but his directional gyro read 275 degrees. He would have to turn if he wanted to locate himself over the station. *Slow, shallow turns!* And then he saw the flag. It read FROM. *Oh, no!* Again he twisted the OBS knob until he got a TO indication and the needle returned to center with a 361-degree reading. *Think, man, think. That means that I'm south of Red Bluff and I'm going to have to turn again, north! What happened to that airspeed? It's down again.* He tried to relax the backpressure on the yoke, tried hard to think which way to turn in order to head north. *And why is that magnetic compass flipping around? Why doesn't it show the same thing as the DG? I've got to make sure the DG is right, but how? Straight and level, that's it. Come on now, straight and level!*

And then turbulence hit full force.

181

I Don't Know Where I Am

In the muted calm of the Red Bluff Flight Service Station the two men were relaxed.

"Whatever happened to Two Whiskey Alpha?"

"Beats me. I called a couple of times, but he didn't answer."

"Think he's in trouble?"

"Maybe yes, maybe no. Didn't sound too relaxed, but then in this kind of weather, a lot of 'em get a bit nervous."

"Why don't you call him again?"

"Sure, why not." His foot pressed on the contact bar, activating the mike. "Mooney Two Whiskey Alpha, this is Red Bluff, do you read Red Bluff?"

Nothing.

"Maybe he's down low and can't hear us."

"Or maybe he's simply turned his radio off or switched frequencies. Mooney Two Whiskey Alpha, do you read Red Bluff, over."

They waited.

"Forget it."

"I guess so." He unwrapped a somewhat soggy ham on rye and bit into it. "Sometimes I think I'll get married again, just so she can make me decent sandwiches."

"You don't know when you're well off."

"Red Bluff! Red Bluff!" The voice which suddenly shouted through the speaker was slurred, trembling.

"Aircraft calling Red Bluff, go ahead." He shoved the sandwich aside. "I bet that's the Mooney."

"This is Mooney Two Whiskey Alpha. I don't know where I am. Can you get me out of here?"

Six miles above the terrain a United Airlines DC-8 was passing over Oakland en route from Los Angeles to Seattle.

"You take it," the captain told the copilot while turning up the sound on the number-three com radio until the voice came in loud and clear.

"Mooney Two Whiskey Alpha, this is Red Bluff. Are you on an IFR flight plan?"

"No, I'm on no flight plan at all. I need . . . ah . . . help."

"Right. Are you VFR now?"

"I haven't got any idea where I am. I've got to get out of this fog. I can't see a thing."

"Are you transponder equipped?"

"Am I what?"

"A transponder, do you have a transponder in the airplane?"

"Oh, transponder. No, I don't have a transponder."

By now others were gradually becoming involved in the drama-in the-making. Not only was the United captain listening in his perch way up above all the weather, but Red Bluff had also contacted Oakland Center, asking them to try and establish radar contact. Oakland, in turn, informed the pilot of an Aztec at 4,000 feet on an IFR flight plan from Sacramento to Red Bluff to be prepared to hold because of an aircraft in trouble.

"What kind of trouble?"

"We don't know exactly. Red Bluff says it's a VFR pilot somewhere in the clouds. He doesn't know where he is."

"Oh lovely! That's all I need. Anybody know his altitude?"

"We're trying to find out. Stand by."

"Will do."

In the cockpit of the Mooney, the pilot had opened all the vents but, still, perspiration was pouring down his face. He kept hearing voices coming at him through the speaker, but he couldn't really comprehend what they wanted from him. *If they'd only stop talking and do something!*

"Mooney Two Whiskey Alpha, Oakland Center wants you to call them on 120.4. They have radar there and they'll be able to help you."

"What was that? I don't understand."

"Oakland Center, they want you to switch your radio to 120.4 and give them a call. One-two-zero-point-four!"

"Oh, right. 120.4. I'll try."

It seemed incredibly difficult to operate the knobs on the radio and he had to keep wiping the perspiration from his eyes to be able to see, but eventually he did manage to switch to the new frequency.

"Oakland . . . ah . . . Center, this is Mooney . . . ah . . . Two . . . ah . . . Whiskey Alpha."

"Two Whiskey Alpha, this is Oakland. What is your present position and altitude?"

"I couldn't hear . . . ah . . . what's that you want me to do?"

"What is your present position and altitude?"

"I don't know . . . I don't know where I am." He tried to read the altimeter through the perspiration. "I think it's like 3,000 feet . . . no, more like 3,500."

"Roger. Climb to and maintain 6,000. I repeat, climb to and maintain 6,000. There are mountains all around you and you're below terrain-clearance altitude. Climb to 6,000."

"Six thousand. Okay, I'll try." He advanced the throttle as far as it would go and tried not to pull too hard on the yoke.

"What's your present heading?"

"I'm flying . . . I can't . . . just a minute . . ." Again he wiped his eyes, trying to clear the blurred image of the DG. "I'm flying thirty degrees . . . yes . . . ah . . . thirty . . . West, that is, west."

"Do you have VOR equipment?"

"Do I have what? I don't know what I've . . . ah . . . got."

"Roger. Remain on this frequency."

The United captain had switched to the Oakland frequency and followed the conversation, wishing that whoever was doing the talking at Oakland would be a bit more patient. His thoughts went back some 20 years and to how scared he himself had been the first few times in real weather.

The Oakland voice came back on. "How much fuel do you have left? Over."

"Ah, what . . ."

"Fuel. Look at your fuel gauges. How much fuel have you got left?"

"Ah . . . about a third, it looks like."

"How much is that in hours?"

"Oh, I don't know. An hour maybe, forty-five minutes."

"Roger."

During the ensuing pause the captain was tempted to pick up the mike and tell that poor soul that with 45 minutes of fuel in his tanks there was plenty of time and just to relax. But he didn't. Another voice might only add to the fellow's confusion.

"I still don't know where I'm at. I just don't know." The voice sounded high-pitched, desperate, as if choken with tears.

Flying into the Mountains

At Oakland Center the usually relaxed atmosphere had changed. They had a problem and they all knew it.

"If he's actually flying thirty degrees like he says, he's flying right into the mountains."

"I know. And three hundred is not much better, and neither, for that matter, is west. We've got to get him turned south, away from the high terrain."

"Try again. At least he seems to have the airplane more or less under control. Maybe, if we can get him to make a turn, we can pick him up on radar."

"Mooney Two Whiskey Alpha, Oakland Center."

"Ah . . . I'm here."

"What is your present heading?"

"My present . . . ah, I'm going on a heading of . . . ah . . . twenty degrees."

"Roger. Is that zero two zero?"

"What heading do you want me to go on?"

"Two Whiskey Alpha, are you on a heading of zero two zero? Over."

"I'm heading at the present time . . . ah . . . zero . . ." There was a burst of static, then nothing.

For some time now the Aztec had patiently been flying a holding pattern, glad that the assigned altitude had turned out to be between layers where he would at least have a chance to see the Mooney if it should, by any chance, come his way. He, too, had been listening to the Oakland Center frequency and when there was no further word from the Mooney he couldn't help wondering if zero would prove to be the last word the pilot would ever say.

"This is . . . ah . . . Whiskey Alpha. I still don't know where I am."

The Aztec pilot breathed a sigh of relief.

"Two Whiskey Alpha," the voice of the Oakland controller. "What is your present heading and altitude?"

Again the Mooney failed to answer.

"Two Whiskey Alpha, Oakland Center. Request your present heading and altitude."

"I'm . . . ah . . . flying at . . . ah . . . twenty-nine degrees."

"Roger."

By now the United captain felt certain that he could do better. He picked up the microphone, switching it to the number-three transmitter. Carefully keeping his voice calm and his speech slow and distinct, he depressed the mike button. "Mooney, Two Whiskey Alpha, this is a United Airlines jet high above you. Could you try and tell me where you think you are?"

"I . . . I have no idea."

"Where did you take off from and how long ago was that?"

"I took off from Seattle. That was about three and a half hours ago, I think." And then, after a pause, "I was going to San Francisco."

The captain, who had flown Mooneys himself, looked at his chart to try and figure out how far the Mooney might have flown at what he estimated its cruising speed to be. As best as he could guess, he should be over the valley in the Red Bluff area all right, and there were high mountains in all directions except to the south. "If that's so, that puts you somewhere right into the vicinity of Red Bluff. In what direction are you flying right now? Can you tell me that?"

"I'm still climbing. I'm over 5,000 feet and I'm still climbing."

At this point the Oakland controller came back on the line. "Roger, Two Whiskey Alpha."

"What do you want me to do? How high do you want me to go here?"

The captain had to assume that Oakland preferred to handle things themselves, so he kept quiet.

"Mooney Two Whiskey Alpha, Oakland Center. The MEA at your approximate position is . . . stand by." There was what seemed to all those listening to be a long pause. "Mooney Two Whiskey Alpha, Oakland Center. Climb to and maintain 8,000."

"This is Two Whiskey Alpha. I think I'm . . . ah . . . above the fog."

"You're above the clouds. Do you have visual contact at all?"

"No, I have no contact . . ."

Tell Me What To Do

For a moment the Mooney pilot saw things. A patch of what seemed like sky, dimly lit outlines of clouds, but it didn't last. *If they'd only tell me what to do!* He pressed the mike button, "I'm at 5,000 and I'm heading south about . . . ah . . ." He leaned forward to better see the DG. "About . . . ah . . . five degrees."

"Mooney Two Whiskey Alpha, say again your present heading. This is Oakland. Say again your present heading."

"I'm . . . I'm heading south, I tell you . . . south, three degrees." He pressed his fist against his eyes, then tried to wipe a non-existent film from the face of the compass. *If only those people down there would stop asking questions! Why don't they just tell me what to do?* Suddenly he glanced at the artificial horizon and reacted with shock, seeing it steeply angled. He tried to level the wings but it seemed to take much more effort than it should have.

"Two Whiskey Alpha, Oakland. I understand you're turning southbound from thirty degrees, is that correct?"

The figure 3 seemed to eerily be swinging back and forth in front of him. "Three . . . three degrees."

"Roger. What is your present, I repeat, you present heading?"

He looked out of the window as if that solid wall of nothing could somehow miraculously provide the answer. "My present heading?" He tried to make sense out of the DG, then the magnetic compass, but the meaning of all those figures suddenly seemed to have escaped him. "I'm . . . I'm going pretty near due east . . . due west . . . three degrees." Totally exhausted he dropped the mike into his lap. "I don't know where I'm going," but no one heard him.

"We're trying to get a fix on you now, over."

Without bothering to pick up the microphone he pleaded into nothingness. "Just tell me what to do, please, just tell me what to do."

At Oakland Center the group around the controller working the wayward Mooney had grown.

"Do you have any idea where that Mooney is or where he is headed?"

"He says he's headed east."

"I thought he said west."

"Yeah, last he said west. I don't think he knows which way he's headed." Exasperated, he picked up the mike once more. "Mooney Two Whiskey Alpha, can you give me your present heading?"

"North, I'm heading north."

One of the men, watching the radar screen, pointed and spoke up. "What about that one, 20 miles south of Red Bluff?"

"Could be. Could be anyone VFR on top or even below. How do we know?"

"How about making him fly some turns?"

"I'll try anything, but the way he sounds, I doubt he knows whether he's turning or not."

From one of the others standing by: "I don't know about turns. If he's hand-flying on instruments we could make him lose control altogether."

The United captain, hearing the knock, released the lock on the cockpit door to admit one of the stewardesses who was bringing coffee.

"Here you are." She handed cups to each of the crew.

"I still have no idea where I'm at." Even the stewardess recognized the desperation in that voice coming over the speaker. "Trouble?"

"Some poor slob down there in the clouds. Oakland's been trying to find him."

Shaking her head she turned to leave just as the Oakland voice came on again. "Two Whiskey Alpha, your compass heading, can you give me your compass heading? You're flying by a compass. You have a compass in front of you. Now if you'll just take a good look at that and tell me what it says."

There was a very long pause and the captain was about to pick up the mike again, when he finally answered.

"I'm flying due west."

"Roger. That's two seven zero compass heading, is that right?"

"I believe that's what it is. It's fog. I'm still in the fog. I'm at 7,000 feet and I'm still in the fog."

"Roger. Can you climb some more? You should break out on top at approximately 9,000 feet."

"Got you. I'm climbing out."

The captain still felt that there might be some help he could give. He pressed the mike button. "Mooney Two Whiskey Alpha, do you read United?"

"I read you . . . but I've no idea what I'm doing."

"Two Whiskey Alpha, Oakland. We believe we now have you in radar contact, approximately 45 miles south southwest of Red Bluff. Can you take a heading of zero eight five for about two minutes, over."

"Zero eight five, I have no idea what that is, to tell you the truth about it. Zero eight five . . . which way is zero eight five?"

"Okay, just be sure to keep your airspeed up. Keep your airspeed up above 100 and keep climbing to about 10,000. Over."

The captain now switched the mike to the radio which was tuned to one of the other center frequencies. "Center, United Three Eight Eight."

"Go ahead, United."

"I believe if you ask him to just take cardinal headings like, uh, east, it'll probably be a little easier for him to understand at this time. Maybe he'll relax a little bit."

"Okay, yes, thanks, United." And then, on the other frequency, "Two Whiskey Alpha, this zero eight five heading is only five degrees off east, so just take up east on your compass, east, over."

"I read you but I don't know what I'm doing. What is it you want me to do?"

At this point the United captain switched his mike back to 120.4.

In the cockpit of the Mooney, the palms of the pilot's hands, the shirt, the yoke, everything seemed soaked in perspiration. *Who are all these voices that keep talking to me?* His mind had simply refused to continue to function.

"Okay now, just relax . . ."

Now, who's that?

"Just relax. This is United. We all get into a spot once in a while. If you just relax your hands on the wheel, just for a second, I think we'll calm down. Take your feet off the rudders and then, uh, just shake your hands for a bit and relax, and then go back to it and just head east which is E on the indicator. East heading and, uh, hold that as steady as you can for a minute or so. Just nice and straight and I think we can calm down quite a bit and accomplish quite a bit. Okay?"

"Okay, I got yah."

"You have a good airplane under you. It's a real good machine and, uh, with just a little help it'll do a real good job for you."

There was something so friendly, so relaxing in that slow, fatherly voice, that he felt calmer already. And, yes, suddenly he did recognize the E on the compass and seemed to succeed in holding it pretty steady.

"I'm going east now, at 9,000."

"Very good, that's very good. Fine. Just hold that now and you'll be doing real good."

The United captain let go of the mike button, trying to figure out what to say next, when the speaker came to life again: "My gas is getting low . . . down below a quarter."

"That's all right. There's still plenty. We all make mistakes. Relax and we'll get you out of this real good."

"United, this is Oakland. Most Mooneys have a reserve of a few gallons after they go on empty, three maybe or so. We show him now 40 miles two zero from Red Bluff."

"Roger. If you could work him down the valley toward Sacramento, there's lots of airports which he can find and land on."

"Affirmative. Mooney Two Whiskey Alpha, this is Oakland again. Continue on your present heading. It'll take another 20 miles to get you away from the mountains."

Here the captain felt that a little more explanation might help. "Two Whiskey Alpha from United. Your present position is probably west of the Sacramento Valley up near Red Bluff and the center is going to take you east over the valley and then drop you down in the valley to the south where you'll have a lot better weather."

"Okay, I got yah. I hold east and keep on climbing out."

"That's right. You keep that east heading now and the people at Oakland will tell you when you're ready to descend and that'll put you in the valley. So you just listen to them and relax a little more. I think once in a while you just take your hands off the wheel and shake 'em a little bit and then go back and it'll be pretty easy for you. I'd set up a cruise now with your mixture leaned out so you can conserve your fuel." In order not to upset or confuse him the captain switched again to one of the other Oakland frequencies. "Oakland, United. I'm afraid I'm going to be out of range pretty quick, so you better take it from here."

"Yes, United, and thanks."

"Good luck."

The Aztec pilot was glad the company was paying for the gas, and he was also glad to be reassured that the Mooney, by now, was apparently at a safe altitude and distance from him. He had just decided to call Oakland to remind them that he was still holding and to suggest that it should now be safe for him to go on to Red Bluff, when the voice of the Mooney pilot literally screamed over the speaker.

"This is . . . ah . . . Whiskey Alpha. We're way out of control!"

Oh, no. Not again!

"Two Whiskey Alpha. This is Oakland. Let go of the controls. Release the controls, just release the controls! Let go of them. Over."

Silent pause.

"Two Whiskey Alpha, don't worry about the airspeed building up. Just relax. The airplane will come out of it on its own. If you're at 7,000 feet or higher you're all right at your present position. There's clearer weather about 20 miles south. As soon as the aircraft recovers, try and take up a southerly heading."

The Aztec pilot shook his head. *If that poor sod has gotten himself into a spin he'll never come out.*

"The tank's empty!"

"If it reads empty you still have three or four gallons. We'll have to take you into Red Bluff. It's enough for that."

Why was it always, in a situation like this, ten seconds could feel like hours?

"Two Whiskey Alpha. Has the airplane righted itself? Can you tell me your airspeed?"

Somehow, the Aztec pilot was certain, now, that he would never hear the voice of the Mooney pilot again, but he was wrong.

"My . . . ah . . . about 110 miles an hour."

"If you have one ten you're all right."

There was another long pause, then, "I'm at 5,500 and at the present time I'm going straight north, thirty-three degrees."

"You're heading toward higher terrain again. Try to turn right . . . right. Make the turn just with your rudder. Just use a little right rudder pressure, that's all, not very much, just a little. Keep your hands off the wheel, don't pull back on it or push forward. Just use a little right rudder." And then, the voice changing to a more businesslike tone, "November Four Eight Seven Five Papa, Oakland."

"Seven Five Pop, go ahead." Well, they hadn't forgotten him after all.

"Seven Five Papa, will you give Two Whiskey Alpha a call? He's about 25 miles north of your present position. See if you can raise him."

"Okay. Mooney Two Whiskey Alpha, this is Aztec Seven Five Pop, do you read me, over."

"I read you. I'm at 6,000 feet, heading . . . this thing won't stay straight."

"Okay, just steady down. I'm at 4,500 in the clear between layers so I'm going to try and find you and take you to Red Bluff. Just try and keep a steady S on your compass and try and keep the airspeed around 110. Pull the throttle back slowly until it shows that you're coming down a little, not too much, keep it at less than 500 feet per minute, and pretty soon you'll break out of the clouds you're in. I'm turning my landing light on so just keep looking for a light."

The pilot had slowed the Aztec to the lowest comfortable speed, and he kept staring into the clouds to the north and above. After a while it seemed right to talk to the Mooney some more. Just as he was about to pick up the mike he caught a red flash in his peripheral vision, then another.

"Two Whiskey Alpha, I think I see you. Do you see a light straight ahead and slightly below your altitude?"

"Yes, yes I do."

"Seven Five Pop, this is Oakland. Do you think you can lead him in?"

"Will try. Two Whiskey Alpha, this is Seven Five Pop. Just come on over toward me and I'll take you to Red Bluff. Oakland, Five Pop, am I cleared for the approach?"

"Seven Five Pop is cleared for the approach to Red Bluff Airport, Runway 33. Good luck."

"Two Whiskey Alpha, I'm flying at just a little over 100 miles an hour. Now you stay close to one side of me where I can see you. You follow me and we'll go into Red Bluff and get down."

"I'm at 4,000 feet and I'm following you."

"Seven Five Papa and Two Whiskey Alpha, Red Bluff weather estimated ceiling 1,800 overcast, visibility three, light rain. Altimeter two niner zero one."

"Thank you, Oakland."

"Your light seems to be disappearing, I'm losing you."

"This is Seven Five Pop. Just keep on flying the way you are. We're going down through some clouds and there'll be times when you can't see me for a few moments, but if you hold her steady the way you are, you'll be all right quite near me when we break out." *I hope!* He fed in a little extra throttle to put himself far enough ahead of the Mooney just to make sure that he wouldn't suddenly run into him.

"I lost you."

"That's all right. Just keep her steady the way you're going. Just relax and keep her steady." *And now just don't mess it up in the last minute!*

"Two Whiskey Alpha. I'm now at 3,000 feet and flying 90 miles an hour."

"That's good, you're doing fine. Just keep the nose down a bit and keep her coming down. I'm at 2,500 and I think I can just about see the lights of Red Bluff."

Not a word for quite a while, then, "This is Two Whiskey Alpha . . ." Just that, nothing else. Then: "I can't hold it steady for some reason."

"Relax. Just steer with the rudder. Keep your hands off the wheel and maybe pull back the manifold pressure to about ten inches or so. Pretty soon you'll be breaking out of the clouds. I'm below in the clear now and I'll wait for you here.

"I'm at 2,500 feet now and about 90 miles an hour. I've got no idea what heading I've got."

"Okay, just close your throttle a little more. You're bound to be in the clear in another moment."

"I'm at about 2,000 feet."

"Okay, you'll be able to see the lights on the ground in a second."

"I see 'em now."

"All right. Do you see the airport at Red Bluff? There's a rotating beacon and the runway lights are up very bright."

"Yes, I do."

Well, hallelujah! "Very good, so now just go ahead and land.

Oakland Center, Seven Five Pop. I think I'll cancel IFR and follow the Mooney in."

"Roger Seven Five Papa. Cancelling IFR, and thanks for your help."

"Any time."

"This is Two Whiskey Alpha. What direction am I supposed to land here?"

"I would say it wouldn't make much difference tonight. Just pick a runway and land on it. I'll give 'em a call and tell 'em you're coming."

"This is Oakland. Red Bluff reports they have him in sight."

"Okay, I guess he got away with it, then."

"Can you see him landing?"

"I'm not sure."

"This is Two Whiskey Alpha. I'm having trouble with my landing gear now."

Oh no, not now! "Is that the kind of Mooney with the gear with the long handle?"

"Yes."

"Well, put the handle all the way down again and then try once more in one quick smooth motion and it'll just slip into the catch under the instrument panel."

"Ah, I got it."

"Good."

"I'm going to land to the north."

"That's just fine."

"This is Oakland. How's he doing?"

"Fine. He's got his landing lights on and . . . okay, he's on the ground and slowing down."

"Then he made it."

"We all did. Don't ask me how. Good night, gentlemen."

The foregoing story is based on fact. Many of the communications, especially the voice of the Mooney pilot during periods of stress, are taken from the transcript of the recordings, which were made at Oakland Center and at the Red Bluff FSS. Aircraft identifications were changed to protect those involved.

The incident is presented here to illustrate what the onset of panic can do to a pilot and, at the same time, to show how much help is available if the pilot will only ask for it

Chapter 27

Westchester County to Boston

Throughout the drive from New York City to the Westchester County Airport the pilot kept looking at the sky, wondering if he'd be able to take off all right. According to the information which he had obtained by telephone, the airport was still IFR but was expected to go VFR some time during the morning. Boston, on the other hand, was VFR but there was some weather coming down from Canada which indicated that it might just turn sour later in the day. And with Boston's Logan Airport being one where special VFR operations were not permitted, he might have difficulty getting in there if he had to wait too long for Westchester County to clear.

When he got to the airport it was still IFR with an 800-foot ceiling and visibility less than one mile which put it below the minimums for special VFR takeoff. He asked about reports on the tops of the overcast and was told that they had been reported at 5,000 feet. He now held a little discussion with himself. Even though he did not hold an IFR ticket, he felt completely confident that he could safely climb to VFR conditions on top. The only trouble was that he thought that he might sound less than professional if he should go ahead and ask for an IFR clearance. On the other hand, if he didn't get out of here soon, he might never be able to get to Boston, and getting there was important.

He Made Up His Mind

Finally, he made up his mind. He called Flight Service and filed an IFR flight plan to VFR conditions on top and then direct to Boston. He then preflighted his airplane, fired up the engine, and called ground control for his clearance, hoping rather desperately that ATC wouldn't foul him up with some kind of complicated instructions.

Pencil and pad in hand, just in case, he waited. He was lucky. The clearance simply read "Cleared as filed" plus a transponder code and a departure-control frequency, both of which he jotted down. Then while taxiing to the end of the runway, he set the transponder to the required code, tuned his number two com to the departure-control frequency, and the number one com to the tower.

It's uncomfortable to sit at the end of the runway when the visibility is such that one can't see the other end. Still, he was familiar with the airport and its surroundings, so he figured that he would be able to take off and climb to a safe altitude without getting into any trouble.

"Ready to go."

"Cleared for takeoff."

He accelerated down the runway, lifted off, and moments later was in the soup.

"Contact departure control now."

He did, was told that he was in radar contact, and continued his climbout by instruments, careful not to take his eyes off the instruments so as not to get bothered or confused by the masses of grey clouds rushing by his windows on either side. It didn't take long, though it seemed as if it did, until things got lighter above and moments later he broke out into brilliant sunshine above a solid undercast. He leveled off at 7,500 feet and called departure control and cancelled IFR.

"Roger, cancelling IFR."

He was on his own and he breathed a sigh of relief. It had been the first time he had actually filed IFR and he had gotten away with it. He now called Hartford for the latest Boston weather and found that it was down to 3,000 broken and three miles in haze. Well, if it stayed that way he'd be able to make it all right. He'd be able to get down VFR through breaks in the overcast and if the visibility held a VFR approach and landing would be legal. But what if it got worse? Well, he'd simply have to wait and see. After all, if all else failed, he had sufficient fuel to get back to New York which, according to the forecasts, was supposed to be VFR by afternoon.

Anyway, it wasn't the first time that he had been VFR on top with no clear idea whether or not he'd be able to get down at his destination. It had always worked before, so why shouldn't it today?

He was tuned to the Putnam VOR which is remoted from Worcester when it was time for the hourly weather sequence, and this time the report for Boston was not what he had hoped to hear. Three thousand overcast, visibility two. Not only wouldn't he be able to get down through the solid overcast, even if he could, he wouldn't be able to land at Logan without an IFR clearance. And one of those in one day was about all he was prepared to handle. What about the other three airports in the area? There were Norwood, Hanscom, and Beverly, and each of them would have to give him a special VFR clearance if he asked for it, assuming, of course, that he'd somehow find a way down through the overcast.

He continued on at the 7,500-foot altitude which would keep him above the Boston TCA and decided to fly to the vicinity of each of those three secondary airports to see if he might not be able to locate some breaks in the overcast. He vaguely remembered that the weather sequence had included a report of broken conditions for Worcester, so he figured that if worse came to worst, he could land there and then rent a car and drive the 35 miles or so into Boston.

It took a while, but everything underneath remained a solid mass of clouds. Apparently, there was no chance in the immediate vicinity of Boston to get down, unless he was willing to simply drop down through the clouds. But he somehow didn't feel comfortable with that idea despite the reported 3,000-foot ceiling. He was certain that he could have made it if he had to, but he just didn't feel comfortable with the thought. So he tuned his nav receivers to the 010-degree radial from Putnam which should take him right on top of the Worcester airport. After a while, he called Worcester for the current weather and was given 1,500 scattered, 4,000 overcast, visibility five. Breaks in the overcast to the north. So that place was going down too. He better hurry up.

The Clouds Were Solid

When he finally arrived over Worcester the clouds below were solid, so he turned north to look for those reported breaks. He'd been heading that way for maybe five or six minutes when, off to his right, he spotted something that looked as if it might be a break. He turned toward it and once he got there he saw something that looked like water and a bit of shore. He checked his chart and decided that what he was looking at must be the Wachusett Reservoir. Fine. Granted, there was a 2,049-foot broadcast tower

Fig. 27-1. Then he was in the clear again ... .

indicated just to the southeast of the reservoir, but as long as he stayed over the water he would be safe.

He trimmed his airplane for a reasonable rate of descent and started to circle down through the open space, constantly keeping the water beneath in view.

Suddenly, he found himself in the clouds, but only for a second or two. Then he was in the clear again and though the visibility wasn't much, he saw that all the clouds were now above him (Fig. 27-1). He leveled off, took up a southwesterly heading, and called Worchester Tower to announce his imminent arrival.

He rented a car, drove to Boston, took care of his business there, stayed overnight, drove back to Worcester the next day in bright sunshine, and flew back legally VFR all the way.

Chapter 28

Running Out of VFR

What follows is a classic case of attempted VFR flight under exceedingly marginal conditions by a crew unprepared for such an operation. The event is related in its entirety in the form of conversation between the pilot and copilot, a transcript of which was provided by the National Transportation Safety Board. Though the airplane involved was, in fact, an airliner, a Convair 600 turboprop, and the crew was instrument rated, it does serve to illustrate dramatically how even experienced pilots can get themselves into trouble. Though there is no record of it, it must be assumed that the decision by the crew to forego the usual IFR clearance and to proceed VFR instead was based on the reports of a line of thunderstorms lying across their route, and the assumption that it would be more expedient and possibly safer to deal with them by staying in VFR conditions, despite the fact that the flight was to take place after dark.

The aircraft was equipped with a cockpit voice recorder which was subsequently retrieved and yielded the conversation which follows. It might be pointed out that indications are that the crew did not have any Sectional or WAC charts for the area in the cockpit and thus had no clear idea of the terrain. One of the many lessons to be learned from this case is that VFR flight under marginal weather conditions should never be attempted under any circumstances unless WAC or, perferably, Sectional charts are available.

Shortly After Takeoff

The cockpit-recorder tape starts shortly after takeoff: "That might not be a hole there." It was the voice of the captain.

The copilot replied, "We'll know shortly. It sorta looks like twenty-four miles to the end. I don't mind, do you?"

"I don't care, just as long as we don't go through it."

"Looks a little strange through there. Looks like something attenuating through there."

"It's a shadow."

"Yeah, looks like a shadow."

"Is that better?"

"Naw, I don't care."

"Suit yourself."

"Well, I don't know, looks a little lighter in here. This thing hits your eyeballs pretty hard."

"That's right."

After a while:

"See something?" The captain.

"I still think it's a shadow." The copilot.

"Yeah, why not."

"All right."

"I'd slow it up a bit, too."

"What have we got, decreasing ground pickup?"

"I didn't hear you."

"The visibility is dropping." Then, "Rain!"

"Raining all over the place."

And some minutes later:

"What's all this. Are those lights in those fields? What are they, chicken farms?"

"Yeah."

"Gosh almighty, they're planning to grow a few eggs, ain't they?"

"That's what they are."

And later still:

"There's not much to that but we've got to stay away from it or we'll be IFR."

"Shoot, I can't get this stupid radar. Got any idea where we're at?"

"Yeah. Two one six will take us right to the VOR."

"Two . . . ah . . ."

"Two oh one I've got."

"Fifteen."

"I'm not concerned with that." This is the captain. "I couldn't care less. I guess you're right. That is just extending on and on as we go along because it hasn't moved in about three or four miles in the last thirty minutes, it seems like. I guess it's building up that way now." After a moment: "What's Hot Springs?"

"Sir?"

"What's Hot Springs VOR? It is ten-zero, is that right?"

"Yeah, yeah, that's right. We don't want to get too far up the . . . it gets hilly."

"Yeah." Then, "Stars are shining. Why don't you try two thousand?" And a few seconds later, "If we get up here anywhere near Hot Springs we get in the mountains."

"Uh, you reckon there's a ridge line along here somewhere? Go down five hundred feet, you can see all kinds of lights. Let's go ahead and try for twenty five hundred."

After apparently having climbed to a somewhat higher altitude:

"All right, Fred, you can quit worrying about the mountains because that'll clear everything over here."

"That's why I wanted to go to twenty-five-hundred feet. That's the Hot Springs highway right here, I think."

"You're about right."

"Texarkana . . . no, it ain't either. Texarkana's back here."

"Texarkana's back over here someplace."

"Yeah, this ain't no Hot Springs highway."

By now we are 18 minutes and 34 seconds into the flight:

"Well, thirty degrees . . . thirty degrees takes you right to Texarkana, doesn't it? Hot Springs . . . here we're sitting on fifty."

"Yeah. How're we doing on the ground?"

"I don't know, Fred. Still keep getting another one popping up every time . . . every time."

"If we keep this up indefinitely we'll be in Tulsa."

"I haven't been in Tulsa in years."

"Ha ha. The last time, I was with Glen Duke. He said, go which-ever way you want to. I was going out of Abilene going to Dallas. Took up a heading of zero one zero and flew for about forty five minutes and he said, 'Fred, you can't keep going on this heading.' I said, 'Why?' He said, 'You're gonna be in Oklahoma pretty soon.' I said, I said, 'I don't care if I'm in Oklahoma.' He said, 'Fair enough.' And then, after a moment, surprised: "How'd I get all this speed?"

"You're all right."

"Pile it on? We'll keep this speed here?"

Fig. 28-1. "There ain't no lights on the ground over there."

"A little while."

"There ain't no lights on the ground over there." (Fig. 28-1).

"Yeah, I see 'em behind us. See stars above us."

"I got some lights on the ground."

"There's just not many out there."

"Maybe . . . Could be something else, coach."

"Ah, we're getting rid of the clouds!" And a few seconds later: "We is in the clouds, Fred."

"Are we?"

"Yeah . . . No, we're not. I can see above us."

"We got ground up ahead?"

"I can see the ground here."

"Yeah, I can see the ground down here too."

"I can see some lights over here."

"That's probably Hot Springs, coach."

"Yep, could be. Yeah, that might be either it or Arkadelphia."

"Well, I'm getting out of the clouds here but then I'm getting right straight into it."

"Oh, looks like you're all right."

"Do you see any stars above us? We're getting in and out of some scud." (Fig. 28-2).

"Yeah, we've got a little bit here."

"I sure wish I knew where we were."

"Well, I'll tell you what, we're ah . . . on the two fifty, two sixty radial from, ah . . . Hot Springs."

Little Success with Radar

The copilot was trying apparently with little success to use the airborne weather radar.

"Painting ridges and everything else, boss, and I'm not familiar with the terrain." Then, "We're staying in the clouds."

"Yeah, I'd stay down. You're right in the base of the clouds. I'll tell you what. We're going to be able to turn here in a minute."

"You wanna go through there?"

"Yeah."

"All right. Good. Looking good, looking good."

"That's all right. Wait a minute."

"Well, I can't even get Texarkana any more."

Fig. 28-2. "We're getting in and out of some scud."

"I'll tell you what, Fred."

"Kay, boss."

"Well, ah . . . we'll just try that. We'll try it. We're gonna be in the rain pretty soon. It's only about two miles wide."

"You tell me where you want me to go."

"Okay, give me a heading of, ah . . . two ninety."

"Two ninety."

"You got six miles to turn."

"We're in it."

"Huh?"

"We're in solid now."

"Are we?"

"Hold it."

"Start your turn . . . standard rate . . . level out and let met see it when you hit two ninety."

"Aw, okay." After a moment, "There's your two ninety."

"Steady on. Should hit in about a half a mile. Should be out of it in about two miles . . . You're in it . . . Are you through it?" A moment later, "Turn thirty left."

"I can see the ground now. There's thirty. Naw, that's thirty five."

"Keep on truckin', just keep on truckin'."

"Well, we must be somewhere in Oklahoma."

"Doing all the good in the world."

"Do you have any idea of what the frequency of the Paris VOR is?"

"Nope, I don't really give a darn. Put, ah . . . about two sixty five heading . . . two sixty five."

"Heading two sixty five."

"Fred, descend to two thousand."

"Two thousand coming up." Then, "Here we are, we're not out of it."

"Let's truck on."

"Right."

"That's all right. You're doing all the good in the world. I thought we'd get, thought it was moving that way on me, only we kinda turned a little bit while you was looking at the map."

"First time I've ever made a mistake in my life."

"I'll be, man, I wish I knew where we were so we'd have some idea of the general terrain around this place."

"I know what it is."

Fig. 28-3. "That highest point around here is about 1,200 feet."

"What?"

"That highest point out here is about twelve hundred feet." (Figure 28-3).

"That right?"

"The whole general area, and then we're not even where that is, I don't believe."

"I'll tell you what, as long as we travel northwest instead of west, I still can't get Paris." A little while later: "We're about to pass over the Page VOR. You know where that is?"

"Yeah."

"All right."

"About a hundred and eighty degrees to Texarkana."

"About one fifty two . . . The minimum en-route altitude around here is forty-four-hund . . . "

At 33 minutes and 41 seconds after liftoff, the sound of impact.

Chapter 29

A Foiling Fog

Regardless of a pilot's ratings and a plane's state-of-the-art instrumentation, fog can render both useless. That curse simultaneously hexed an IFR student and his instructor. It started one Friday evening . . .

The instructor and his wife were in a Piper Arrow that taxied out in front of the Grumman American Cheetah pilot who was solo. Both were headed northeast; the Arrow flying 200 miles farther than the Cheetah. The Arrow had filed IFR, and the Cheetah VFR. It was already dark when they took off into a crystal clear autumn night.

The Cheetah would fly direct to a VOR as the Arrow deviated to intercept a victor airway. The Cheetah pilot monitored en route air traffic control (ATC), listening to his instructor's technique and lost contact with the Arrow near the VOR. Both made it to the respective destinations without incident.

Sunday the Cheetah pilot had a bad case of "get-home-itis" and despite poor weather conditions decided to try it, hoping the weather would improve after takeoff. Destination weather was poor but VFR so he decided to try it anyway, and turn back if necessary.

A thick but high overcast hung over the airport with scattered clouds underneath. Looking south in the direction of flight, the pilot could see large areas with no clouds under the overcast and visibility was good, so he decided to leave.

Fifteen minutes after departure he was ready to turn back. The scattered clouds he was flying over had gathered and were now broken, so he descended below them . . . and they promptly turned into a second overcast. Visibility was deteriorating and a 180-degree turn seemed likely. But suddenly the visibility improved and he decided to continue, now planning to stop halfway and spend the night.

Getting Down

Stopping for the night was an exercise in itself. Scud clouds surrounded the city and rain with drizzle had started falling at the airport. Radar vectors from approach control included deviations to avoid clouds. Finally the Cheetah's tires touched down on wet pavement that, as it turned out, would not be dry anytime soon.

The weather quickly deteriorated to a steady rain with some heavy showers. Safe and dry in a motel room, he fell asleep to the sounds of rain on the roof, dreaming of sunshine and calm winds Monday morning. It dawned gray, wet and cold with some fog. Trying to get some work done at the airport he decided to check bus schedules and learned the only bus going his direction had just departed. If the weather did not improve he decided to be on that bus Tuesday, losing just a day and a half of work.

It got worse Monday afternoon as a thick fog settled over the area disrupting all civil air travel. On the bus Tuesday, fog was so thick visibility seemed to average about 100 yards. And it stayed that way . . . Wednesday . . . Thursday . . . and Friday. The uncontrolled home airport was closed for several days.

Second Verse

The Cheetah pilot wondered if his instructor had returned in the Arrow before the poor weather settled in. Sure enough his truck was at the main hangar while he probably enjoyed a "grounded" cup of coffee. He made it home under similar circumstances, but this time the Arrow pilot had eavesdropped on the Cheetah pilot.

While the Cheetah pilot was VFR and landing to sit out the weather, the Arrow pilot was IFR above the clouds, and his student. While monitoring an approach frequency, the instructor heard his student getting vectored to the airport. The instructor considered landing to give his student a ride home and tried to reach the Cheetah on a unicom frequency. But that proved fruitless because the airport's unicom was 122.7 and the instructor was transmitting on the standard 122.8.

So the instructor continued, but the fog was settling in, forcing the home airport below the minimums of the NDB approach. He flew to a large en route airport with an ILS and shot the approach. By the time a friend arrived to take the instructor home, that airport was below minimums, too.

Both instructor and student were grounded by a cool and dense fog—the Arrow was 100 miles away and the Cheetah was 200 miles away, both waiting for the weather to clear.

It was a rare instance where a pilot with all the ratings and an airplane equipped with state-of-the-art navigation gear could not get in the air. Fog ruled supreme.

Finally, the Arrow came home Saturday, six days late. It would be used Sunday to retrieve the Cheetah and it, too, got home . . . one week late.

Chapter 30

Postscript

Many a novice pilot, reading this book, will probably say to himself with complete conviction that no matter how much and how long he flies he will never get himself into the kinds of situations which have been described in the preceding pages. And, for a while at least, he may actually succeed in avoiding all those marginal conditions. But gradually, as time begins to pile up in his logbook, his attitude toward what he considers marginal will inexorably change. He will learn to accept and deal with gradually increasing crosswind components during takeoffs and landings. He will take off and fly when it's raining and when the overcast is just barely above VFR minimums, and he will climb to VFR conditions on top when he finds breaks in the clouds which permit him to maintain visual contact with the ground. And eventually he will find himself confronted with the kind of condition which he had earlier on decided to avoid at all cost.

The fact is, if we want to make sufficient use of the airplane to justify the investment in flight training and the cost of owning and maintaining all or a piece of an airplane, we have to do a lot of flying, and that just isn't possible if we insist on restricting our aviation activity to CAVU days.

An Instrument Rating?

The question which naturally comes to mind is, should we get an instrument rating? The answer must necessarily be a qualified yes. The trouble is that instrument ratings are not only expensive but also difficult to obtain, and once a pilot is instrument rated, it takes a lot of continuing practice to remain sufficiently proficient to be able to safely use it when the need arises.

Different pilots have different attitudes toward instrument ratings. There are those, mostly old-timers, who simply feel, who needs it? They started flying in the days when the airspace was still free, when the concept of air traffic control was still a gleam in the eye of some executive of one of the infant airlines. They were, and those who are still around probably still are, excellent pilots. But in terms of today's utilization of the airspace, they must be considered Neanderthals, relics who don't fit in any longer.

Then there are those who simply cannot afford the cost and/or time involved in going through the rigorous training involved with obtaining an instrument rating. They usually keep saying to themselves that they'll do it one of these days, but time passes and they continue to successfully squeak by, staying more or less VFR, and the urgency and desire to be instrument rated gradually recede, until it becomes one of those things you always wanted to do but never got around to.

Another category is the pilot who actually took all the required hours of instrument training but who then flunked the written test. The instrument written is difficult and much of it actually consists of subjects which prove to be of little value later on. Most of us, once our school and college years are behind us by a decade or so, find it extremely difficult to readjust our attitude and thinking toward cramming for an exam. As a result, once having flunked the test, we may be tempted to decide that we know enough about instrument flying in order to use it when the occasion arises and that, therefore, it isn't worth the effort to try and take the test again.

Two Basic Categories

Instrument-rated pilots, too, fall into two basic categories. There are those who obtained their rating and who use it constantly. They will fly instruments even when the weather is VFR and virtually every approach to a landing becomes a practice instrument approach. These pilots become and remain proficient and they learn to recognize their own limitations as pilots and the limitations in terms of performance and available equipment of the aircraft they are flying. But, excluding the professionals flying for the airlines, corporations or air-charter services, their number is most probably in the minority.

A vast number of instrument-rated pilots consists of businessmen who use their aircraft in their business or profession. Initially, after they obtained their instrument rating, they may have used it with some degree of frequency. But as time passed, they found that more often than not it is simpler and quicker to operate VFR. In consequence, they gradually used the instrument ticket less and less. When they did use it, it was probably just to climb through an overcast to VFR conditions on top or to descend through an undercast to an airport where the ceiling was 1,000 feet or better and the visibility ample. But then there always comes that day when one simply feels that one has to be at a certain place though the conditions are barely above IFR minimums. Then the psychological attitude which results from having an instrument rating and therefore being in a position to legally make the approach, will influence the pilot and he will try to make that VOR or ADF approach to that airport where he has never been before. Most of the time he'll probably make it, but then there may always be that one time when he doesn't.

The trouble is that for the average general aviation pilot making non-precision and even precision approaches under minimum or near-minimum conditions, it is a lot harder than it is for our airline counterparts. They operate with two- or three-pilot crews, flying aircraft equipped with every conceivable electronic aid (most of them in duplicate or triplicate) and they nearly always fly into airports where they have been hundreds of times before. We, on the other hand, are usually alone in the cockpit, often fly to unfamiliar airports, and must deal with much less sophisticated instrumentation. There are those who claim, with considerable justification, that no single pilot should attempt to fly a tight instrument approach in an airplane which is not equipped with a reliable autopilot. The pilot workload involved in communicating with ATC, studying approach charts, and at the same time flying the airplane, usually in turbulence, is just too much to be handled safely by one person.

The question still remains: Should every serious pilot attempt to become instrument rated? There is no hard and fast answer. One consideration is money. If finances permit undertaking the expense not only of the flight training involved, but also of obtaining the kind of instrumentation for the airplane which will turn it into an efficient and safe instrument ship, then, by all means, the rating should be obtained. There can be no question that instrument training makes us better pilots, as long as we continue to be careful not to fall prey to overconfidence. This type of training, more than any other, teaches us to always stay ahead of the airplane, to be comfortable with it when it and its instruments are the only thing that stand between life and death.

The Written Exam

And if passing the written exam seems too troublesome, there are a number of organizations around the country which specialize in cram courses. These courses usually involve just one weekend of continuous lectures and study, culminating in the exam on Monday. Most of these organizations will guarantee that you pass the exam, meaning that if you fail, you can retake the course and the exam free of charge for as many times as is necessary to eventually pass. It should be said here that these courses don't actually teach you much. All they are designed to do is get the student over the hurdle of that pesky written. Pilots contemplating obtaining an instrument rating might be best advised to take one of these cram courses first and thus to get the written out of the way. They can then clear their heads of all that extraneous junk that is part of the written and concentrate on the flight portion of the training, which is where actual instrument flying is being learned.

If money is a problem, if the best you can do is take an instrument flight lesson every week or two, it may not be worth it at all. Instrument training requires a great deal of concentration. There is much to be learned and to be remembered, and when such training is squeezed in between the day-to-day routine of running a business or profession or of efficiently performing some type of job, there may not be enough energy left to absorb all of this information which the instructor will be trying to convey.

The number of VFR pilots will always exceed the number of those with instrument ratings. It's nothing to be embarrassed about. As long as we remain aware of the limitations of our capabilities and the performance limitations of the airplanes we fly, we will be able to operate safely most of the time. Let's never forget that, regardless of which part of the country we are talking about, with the possible exception of the Aleutians and parts of Alaska, the weather is good much more often than it is bad. If we're willing to accept the fact that there are times when we simply can't fly, or can't get to where we wanted to go, then there is no reason why the lack of an instrument rating should present any serious problem.

Glossary

ADF—automatic direction finder.

agl—above ground level.

airport traffic area—an area, 10 miles in diameter, surrounding a controlled airport, including the airspace from the ground up to but not including 3,000 feet agl. When operating within this airspace, aircraft must maintain radio contact with the control tower.

ambiguity meter—TO/FROM indicator in an OBI.

APU—auxiliary power unit.

artificial horizon—an air-data instrument which shows the relation of the attitude of the aircraft to the horizon.

ATC—air-traffic control.

ATIS—automatic terminal information service; recorded information about airport conditions, broadcast continuously.

C—degrees centigrade (or Celsius).

CAVU—ceiling and visibility unlimited.

CDI—course-deviation indicator.

CDU—control-display unit.

ceiling—broken or overcast cloud cover, obscuring 60 percent or more of the sky.

com radio—communication radio.

control-display unit—the panel-mounted instrument which provides the interface between the pilot and the area-navigation computer.

212

controlled airport—any airport with an operating control tower.

controlled airspace—airspace within which IFR traffic must maintain contact with ATC.

control zone—the area around a controlled airport, normally approximately circular in shape and extending five miles in all directions, plus any extensions necessary for instrument approaches and departures, under active ATC control. It extends upward from the ground to the base of the Continental Control Area, or, where not underlying the Continental Control Area, it has no upper limit.

DG—directional gyro.

directional gyro—a gyroscopic compass which must be reset intermittently to conform with the magnetic compass.

DME—distance measuring equipment.

EAA—Experimental Aircraft Association.

EGT—exhaust-gas-temperature gauge.

E6b—a circular flight computer, similar in operation to a slide-rule.

ETA—estimated time of arrival.

F—degrees Fahrenheit.

FARs—Federal Aviation Regulations.

FBO—fixed-base operator.

fpm—feet per minute; the rate of climb or descent.

FSS—Flight Service Station.

HSI—horizontal situation indicator.

IAS—indicated airspeed.

ident—the capability of a transponder to identify a given aircraft on the controller's radar screen when the pilot pushes the ident button on the transponder.

IFR—instrument flight rules. The rules governing operation in weather conditions below VFR minimums.

ILS—instrument landing system.

Jeppesen charts—radio-facility charts produced and marketed by Jeppesen-Sanderson.

kHz—kilo Hertz.

knots—nautical miles per hour.

localizer—the horizontal guidance portion of an instrument landing system.

LOM—a compass locator co-located with the outer marker.

Loran C—A low frequency navigation system utilizing specialized ground-based transmitters to determine an airplane's position in terms of longitude and latitude.

MEA—minimum en-route altitude.

MHz—mega Hertz.

mph—statute miles per hour.

msl—height above mean sea level, in feet.

nav aid—a ground-based station equipped with electronic navigation aids, such as a VOR or NDB.

nav receiver—radio receiver designed to receive signals from ground-based nav aids.

NDB—non-directional beacon.

nm—nautical mile(s).

Notam—notice to airmen; notices published periodically by the FAA.

OBI—omni-bearing indicator; a panel-mounted instrument which displays the information received by the VHF nav receiver.

OBS—omni-bearing selector; the knob on the OBI which permits the pilot to select a given radial or bearing from or to a station.

Pilotage—flying by reference to visual ground-based landmarks alone.

RNAV—area navigation.

rpm—revolutions per minute.

Sectional—an aviation chart which includes, among other information, terrain features. It is the preferred type of chart for VFR flight. Scale: 1 to 500,000.

service ceiling—the altitude at which an aircraft, at full gross, will still climb at 100 fpm, under standard atmospheric conditions.

sm—statute mile.

SVFR—special VFR.

TAS—true airspeed.

transponder—a panel-mounted electronic pulse instrument which enhances the aircraft's radar return on the controller's radar screen.

transponder code—the frequency channel to which the transponder is tuned. ATC will request that the pilot tune his transponder to a given code with the phrase: *Squawk zero one hundred* (or some such number).

turn-and-bank indicator—also known as needle-and-ball; a simple cockpit instrument showing the attitude of the wings relative to the horizontal plane, and the correct use of rudder during turns.

uncontrolled airspace—airspace in which VFR minimums are less than those in controlled airspace, and in which IFR traffic may operate under instrument conditions without being in contact with ATC. ATC has no jurisdiction over uncontrolled airspace.

Unicom—a radio transmit-and-receive unit at uncontrolled airports or FBOs used to transmit or obtain unofficial information.

Vertical-speed indicator—an air-data instrument showing the rate of climb or descent.

VFR—visual flight rules; or the weather conditions under which VFR traffic may operate.

VHF—very high frequency.

Victor airways—airways in the airspace below 18,000 feet msl, which are referred to VORs.

VOR—very high frequency omni-directional radio range; a ground-based nav aid, the primary means of radio navigation in the U.S.

VORDME—a VOR co-located with a DME facility.

VORTAC—a VOR co-located with a TACAN facility. For all practical purposes the same as a VORDME.

VSI—vertical-speed indicator.

WACS—World Aeronautical Charts; similar to Sectionals but in a scale of 1 to 1,000,000.

waypoint—a phantom VOR artificially created through the use of RNAV equipment.

W/P or WP—waypoint.

yoke—control wheel.

Zulu time—Greenwich Mean Time.

Index